The Project Manager's Guide to Purchasing

In memory of Graham Ritchie.
Not the book we had intended, but a start.

The Project Manager's Guide to Purchasing

Contracting for Goods and Services

GARTH WARD

Routledge
Taylor & Francis Group

LONDON AND NEW YORK

First published in paperback 2024

First published 2008 by Gower Publishing

Published 2016 by Routledge
4 Park Square, Milton Park, Abingdon, Oxon OX14 4RN

and by Routledge
605 Third Avenue, New York, NY 10158

Routledge is an imprint of the Taylor & Francis Group, an informa business

Publisher's Note
The publisher has gone to great lengths to ensure the quality of this reprint but points out that some imperfections in the original copies may be apparent.

British Library Cataloguing in Publication Data
Ward, Garth
  The project manager's guide to purchasing : contracting for
  goods and services
  1. Contracting out 2. Purchasing 3. Contracts 4. Project
  management
  I. Title
658.7'23

Library of Congress Control Number: 2008921308

ISBN 13: 978-0-566-08692-2 (hbk)
ISBN 13: 978-1-03-283776-5 (pbk)
ISBN 13: 978-1-315-55385-6 (ebk)

DOI: 10.4324/9781315553856

Contents

List of Figures

List of Tables

Acknowledgements

Firstly, I must thank my friend and colleague David Wright who has allowed me to use two key exercises (*Fly Me To The Moon* and *Which Would You Pick?*) both of which are central to the theme of the book. Over the years David has educated me about the subtleties of contract law. Further, he has been invaluable in allowing me to pick his brains and use him as a sounding board for new ideas.

Two other individuals, Dr Roy Whittaker (Project Group Manager of the old ICI – and one of my project management heroes) and Vernon Evenson (friend and colleague and Project Manager with leading contractors and clients), deserve equal thanks. They took time to write to me detailing their experiences and provided thoughtful responses to my questions. Where appropriate, specific extracts have been identified. However, much of their wisdom has been subsumed into the body of the text.

Secondly, I must express my gratitude to two people; Steve Davies, Managing Director of Foster Wheeler Energy Ltd and Mike Cleaver, Commercial Vice President of M W Kellogg Ltd, who arranged for their organizations to provide me with forms and documentation. Steve and Mike made sure that I got exactly what I wanted – many thanks. My appreciation of the support provided by Foster Wheeler goes back over many years. They were always willing to provide guest speakers, in procurement and materials management, during my 10 years as Director of the MSc in Project Management at Cranfield School of Management, and I am sure that what I learnt from them is reflected in what I have written.

There are two other groups of people that I must acknowledge. Firstly, all those who came to Cranfield as guest speakers. They were all recognized as pre-eminent in their respective fields and over my time at Cranfield they provided me with a superb education. Two of them, who have allowed me to use extracts from their material, and who, thus, deserve special mention are Tim Blackford and Nigel Parry. I met both of them when I was Director of the EITB and later ECITB (Engineering Construction Industry Training Board) 7-day course on Project Management for Engineering Contractors. Later they both came to Cranfield. Tim taught me that the principles were the same in all business and industry sectors having worked in senior procurement positions as client, contractor and subcontractor in defence, power generation, petrochemical, paper, and rail. Similarly, Nigel provided insight into 'how clients choose contractors', from his time in project management at ESSO, and how it applied equally to the public sector from his time as a consultant to different government departments.

Secondly, all the people quoted in this work have my gratitude. Similarly, I am very grateful for all the conversations with numerous people, in different positions, in the projects business. If there is anyone who thinks that I have failed to acknowledge any material that I have used, I will gladly do so in the next edition of this book.

My thanks also go to my Editor, Jonathan Norman, who has always provided good advice at the right time, and Fiona Martin, the Assistant Editor, who helped me resolve numerous problems along the way. Also, a big thank you to other members of the team, in particular Gillian Steadman and Charlotte Parkins, who were so considerate of my wishes.

Finally, as in all award ceremonies I must thank my family. My three sons Gavin, Giles and Guy all contributed in some way and commented on the chapters that they were *told* to read! However, the Oscar goes to my wife Gwyneth who corrected my English and did the proof reading, even though she *hated* doing it!

Without the contributions from all of these people this book would be the poorer.

Preface

This book is dedicated to my friend and colleague, the late Graham Ritchie, remembering the days when we were both Project Managers at Bechtel. At that time Graham started writing a book on project management. He soon realized, however, that it was a bigger task than he had anticipated and he asked me to be his co-author. We intended it to be the most comprehensive book on the subject ever written, and assembled a substantial portion of the necessary material. However, Graham left to take up the post of Director of the MSc in Project Management at Cranfield School of Management – one of the world's top ten business schools during Professor Leo Murray's leadership as Head of the School.

In the course of Graham's short time as Director of the MSc he produced a staggering quantity of project management teaching material. It saddens me that so much of his work is no longer in use. However, one of his notes on procurement forms the basis of the early part of Chapter 2.

At the same time, I was leading a team that had put its heart and soul into a proposal and lost. As a result, the company offered me the job of manager of a function that I really did not want to do. In the absence of any other project work, I said yes. Consequently, when Graham asked me, a short time later, to join him in developing a consultancy and training business, I also said yes. Soon afterwards he was diagnosed with a brain tumour and Cranfield asked Bechtel to loan me to them as Acting Director of the MSc Course in Project Management.

I genuinely believed that Graham would be cured, but 8 months later he died suddenly and I took over his role as Course Director. Ever since, though, I have traded under the name of Ritchie Ward Associates, since it was his idea.

Over the years I have tended to focus on commercial subjects, since I have found them to be the least well understood in the project management world. I hope that this book, which represents a part of what Graham started 20 years ago, adds to project management expertise.

Garth Ward

1 *Introduction*

Purchasing is not new; it has been around since the invention of money, and it is even older if we include bartering as its forerunner. Consequently, we are all familiar with it. Further, we all buy goods and services if we consider the domestic as well as the business context. However, we may not be fully aware of the contracting processes involved. Fortunately, in the domestic environment we are protected by the 'Sale of Goods' act. In the business context the game is different. Companies are considered as being able to look after themselves, put procedures in place and have professional advisors. The Romans had a saying, 'caveat emptor' (let the buyer beware) which is still part of our legal philosophy. As a consequence, organizations are careful to ensure that trained personnel carry out buying. Further, the purchasing specialists have been 'authorized' by the organization to form contracts on their behalf. However, even then, they may not be fully aware of the implications of the contracts involved.

This book is written from a project manager's perspective but it is not a book on project management as such. Since buying goods and services is an integral part of most project processes, there will be many instances where project management methods will be discussed. However, a specific point of view will be taken, and detailed debate over project management philosophies will be avoided. Further, the procurement manager on a project is, after all, the manager of a portfolio of smaller projects. Consequently, understanding the context in which purchasing takes place is essential if the process is to be managed effectively.

The book is intended for those who wish to improve their knowledge of the purchasing process from a project perspective. After being involved in education and training for 20 years I have concluded that you can never be sure where people lack the necessary knowledge or experience. Consequently, the book assumes little prior knowledge. However, it endeavours to extend the knowledge of experienced buyers and other project personnel.

When purchasing goods and/or services one is involved in a contracting process, and this is the focus of the book. However, the process involved in the suppliers' or contractors' tendering phase is not covered in detail. The intent is to focus on principles rather than the particular peculiarities of individual business or industry sectors. Nevertheless, some references to proprietary processes will be used to demonstrate the extent to which the principles are varied.

Every effort has been made to make each chapter stand, or to be read, alone so that they are not dependent on knowledge from other chapters. Nevertheless, a reference to the appropriate chapter has been included where a topic is integral to the subject under discussion but is covered in more detail elsewhere. Occasionally, some phrases are reiterated where the issue is of importance to the subject under discussion.

Each chapter deals with the general before discussing the particular, and deals with both the purchasing of goods and services. The buying of goods is used as the 'generic' element applicable to both aspects. Contracting for services is used to demonstrate the more complex issues involved.

Experiences and examples have been included to demonstrate the diversity of issues that the project manager can be involved with, but also to transpose the theory into real life.

A distinction is made between Procurement and Purchasing. Purchasing is restricted to the act of buying the goods and services; whereas procurement is managing the whole process from buying to delivery of the goods.

A term that is used more and more these days, in project organizations, to distance the procurement function from the more narrow purchasing process is Materials Management. It does focus more on what a project needs, but it is to all intents and purposes, procurement with material control and inventory control added on.

A section explaining additional terminology used appears at the end of this chapter.

PROCUREMENT

A procurement department usually consists of four or five groups:

Purchasing:	responsible for buying the materials and equipment.
Expediting:	responsible for ensuring that the goods are ready on the dates required.
Inspection:	responsible for ensuring that the goods meet the desired quality and specifications.
Shipping/Traffic/ Transport:	responsible for ensuring that the material is transported from the manufacturers' or suppliers' works to the project location.

A contracts group responsible for formulating the subcontracts for services, for execution and administration by the installation or construction department, is also required. In some organizations this group is part of procurement, for example, for formulating orders for design services. In some a separate contracts group is established, and in others the formulation of contracts is part of the executing department, for example, for installation and construction services.

The purchasing of 'goods' comprises both materials and smaller equipment, and will involve placing orders using a company's standard contract terms. Large equipment and packages will, more than likely, involve installation and commissioning services. Contracting for design, installation or construction services often involves the provision of materials as well as services. As a consequence, they will involve more specialized and detailed contracts.

A procurement department or group on a project will be headed by a project procurement manager, supervisor or coordinator. The project procurement manager's role is to coordinate purchasing activities and the interfaces with the other functions within the group. They are also responsible for coordinating with other disciplines such as design, installation, accounting and legal departments. They implement the procurement plan within the overall project execution plan and control the procurement schedules. They ensure that decisive action is taken should problems arise and maintain relationships with the owner.

Whilst there is a scheduled procurement phase in a project during which the bulk of the purchasing of goods and services takes place, the procurement function should also be involved in the project development process.

In the early feasibility stages of a project, the procurement function should be providing the owner's project team with information and advice about the marketplace for the key (long lead) equipment critical to the success of the project. Later in the planning and basic design phases, detailed information concerning likely sources of supply, availability, costs, delivery times and foreign currency requirements for the key equipment will be validated. At this stage the information will be expanded to cover other specialized items and bulk materials. The procurement department will also be investigating the capabilities and workloads of potential contractors for the project. They will be developing a recommended tender list for all goods and services, and recommending appropriate contract strategies for discussion with the project manager and other members of the project team. The appropriate specialists in the department will be investigating transport routes, restrictions and constraints, methods of transport, permits and formalities required, customs clearance procedures, duties and so on. They will be checking loading and discharge capabilities as well as handling facilities at ports and transhipment locations.

Once a client has awarded a contract to a contractor to execute the project, the project procurement manager assigned to the project team prepares procedures based on the contractor's corporate procedures. The procedures will be tailored to suit the client's requirements, the specifics of the project and project management's strategy and objectives.

The project procurement manager is responsible for the issue of a multitude of status reports and will be involved in reviewing them, on a regular basis, with the project manager. During the execution phases of a project the project procurement function generates more paperwork than any other department. The following are the main generators of this paperwork:

- the buyer's report;
- the material status reports;
- vendor data and drawings.

The buyer's report lists each group of materials or equipment for which enquiries will be issued, listing the names of all the companies who have been invited to tender. The enquiry issue date and the tender due date are shown and, as each tender is received, the date and time of receipt is noted. When the order is eventually placed, the successful vendor is indicated in the report.

The material status report is a much more detailed report which lists each purchase order once it has been placed. The report then includes a complete history of the order until it is actually delivered. Input for the report comes from expeditors (desk-based and field-based), inspectors and shipping specialists. The report will show the required delivery date and the forecast delivery date.

It is the design department that generates the requests for the vast majority of vendor information and this documentation is often controlled by a separate group.

Procurement status reports or materials management systems are crucial to the project manager's ability to control a project. The systems must be sufficiently flexible to allow one to select different fields of data to produce different reports, for example, by supplier, commodity, material or equipment identification number, work area, drawing number or by promised delivery date. It seems that every company has its own favourite names and formats for each set of reports.

Objectives

Many articles summarize the objectives of procurement as obtaining:

- the right goods and services;
- in the right quantities;
- to the right quality;
- at the right time;
- to the right destination;
- at the right price;
- from the right supplier.

These objectives have the same conflict as the classical Cost–Time–Quality triangle for the project as a whole, and are naturally influenced by the cost, time and quality objectives of the project.

The buying of materials and equipment for a project involving the design and construction of a plant, facility or system is different to the buying of materials for a project involving the manufacture of equipment. In manufacturing, purchasing is more volume-based for repetitive materials. Further, the cost of materials bought from suppliers can make the difference between selling the item and making a profit, or failure to make a sale.

Listen to AT&T's executive vice president for telephone products:

'Purchasing is by far the largest single function at AT&T. Nothing we do is more important.

Simple fact: when the goal is boosting profits by dramatically lowering costs, a business should look first to what it buys. On average, manufacturers shell out 55 cents of each dollar of revenues on goods and services, from raw materials to overnight mail. By contrast, labor seldom exceeds 6% of sales, overhead 3%. So purchasing exerts far greater leverage on earnings than anything else. By shrinking the bill 5%, a typical manufacture adds almost 3% to net profits.[1]

In process plant projects, for example, purchases are more of a one-off event customized to suit the owner's or contractor's specifications. Projects have always striven to buy the right materials, once, at the right time (Just in Time - JIT), in the right quantities, to the right destination. The issues of quality and price can be influenced by who is purchasing the materials – the owner or a contractor. If a contractor is purchasing the materials then they can be influenced by the type of contract that they have with their client. There can be other conflicts in objectives. The owner may have received approval for their project on the basis of the amount of work that the project will generate in the local area. Consequently, they will be seeking to maximize the amount of work done on site. A contractor, on the other hand, will almost always be seeking to minimize the work on site.

Approximately 40 per cent of the total cost of a major engineering and construction project is spent on the procurement of materials and equipment – for a civil engineering project the materials can be as high as 65 per cent. It is, therefore, an extremely important area where it is possible to make significant savings in cost and schedule. Consequently, purchasing objectives can be cost reduction objectives or profit-making objectives. However, in addition to these, there are wider objectives of controlling financial commitments, controlling negotiations and providing information to management.

The design department is responsible for specifying the goods correctly, but, in making the choice of the right materials, different criteria and the needs of different functions should be considered.

Specifying a material that is only marginally better than another, for the service conditions, may require longer delivery and upset the sequence of planned installations operations. Nevertheless, purchasing may *not* revise any technical requirement.

Purchasing can cause a similar result by selecting a vendor's price that is lower, but for which a longer delivery period is required, On the other hand, the installation group may try to finish a job sooner in order to save on overhead type costs, without considering the extra costs that will be required for overtime in the design office, or for the premium freight costs needed to expedite deliveries. Savings in one area are only meaningful if larger expenditures are not required in others.

The right quantity for equipment is easy and straightforward, but the right quantities for bulk materials is something that few, if any, organizations can get right – despite the use of computer-aided design. Prefabricating or modularizing, in multi-locations, compounds the problem of getting the location and quantities correct.

---

1   'Purchasing's new muscle' by Shawn Tully. *Fortune*, February 20 1995.

The placement of purchase orders in the right order and at the right time is crucial to the progress of a project overall. It is not uncommon for buyers under pressure to make progress in purchasing activities by placing the easy orders first leaving the difficult, but critical, ones until later. The placement of purchase orders releases a number of crucial actions. The first is the reservation of manufacturing capacity. This, in turn, starts the delivery process for the fabrication, construction or installation phase. However, before delivery there is a requirement to provide information and drawings for progressing the design or engineering of the facility concerned.

Sourcing from the right supplier may be the most important objective. Choosing a reliable supplier whose quality and delivery performance can be relied upon will save time, effort and money in expediting and inspection. Thus, sourcing at a price may be the least important.

PURCHASING

Purchasing comprises a number of activities for which it has full responsibility, together with aspects for which it shares responsibility with other functions. In addition, it has an interest in some activities that are the responsibility of others.

Purchasing activities for which it can have full responsibility:

- selecting, assessing and rating suppliers;
- issuing enquiries, receiving and evaluating tenders;
- obtaining prices;
- awarding purchase orders;
- following up on delivery promises;
- adjusting and settling complaints and claims;
- developing and maintaining supplier relationships.

Activities for which responsibility can be shared with other functions:

- obtaining technical information and advice;
- obtaining material and equipment costs for estimating purposes;
- contributing to tenders for the sale of goods and or services;
- developing and establishing specifications;
- determination as to whether to make or buy;
- formulating and originating contracts for administration by others;
- responding to questions from suppliers;
- evaluating and settling claims;
- scheduling and timing of orders;
- issuing status reports;
- considerations of quantities and number of deliveries;
- specifying delivery method and routing;
- inspection and expediting;
- transportation, shipping and traffic;
- customs clearance;
- inventory and warehousing control;
- sale of scrap, salvage and surplus;
- forward buying and hedging;
- market research;
- invoice approval;
- purchasing for employees.

Activities of interest to purchasing but the responsibility of others:

*   receiving and warehousing;
*   payment of invoices;
*   administration of contracts for services.

The above listings are not definitive or exhaustive. Each industry or business sector will have its own custom and practice, and specific companies and projects will have their own requirements.

It is difficult for purchasing to be accused of being wrong in what they do. They only buy what they are told to buy by the design department, and the execution/construction or fabrication/installation people tell them when equipment and materials are required. Further, their enquiry and tendering cycle tends to be fixed for each particular type of goods or services. Thus, all that purchasing can get wrong is the choice of suppliers and the contracting strategy to be used. However, even the choice of who is on the tender list and the type of contract has elements of joint decision making. Accordingly, purchasing's main responsibility focuses on administering the process, making sure that key milestones in the cycle are met and that the paperwork is correct. Perhaps their greatest contribution is in obtaining the best deal.

International purchasing

The basic purchasing processes do not change to any significant extent when purchasing internationally. However, a number of aspects will take on greater importance:

*   cultural and language issues;
*   conformance to technical specifications;
*   national standards and codes of practice;
*   legal and tax differences;
*   insurance;
*   logistical arrangements;
*   shipping terms – customs clearance and duties;
*   payment in foreign currencies;
*   letters of credit;
*   bonds and guarantees;
*   the use of agents both for purchasing and freight forwarding;
*   the cost of transport, expediting and inspection.

Outsourcing

New terms describing specialist services now being used are: Outsourcing, Business Process Outsourcing and Offshoring. Outsourcing is delegating the provision of products or services to a third party. Business process outsourcing is handing an entire business process, such as procurement, to a third party. Finally, offshoring is when a company relocates processes, such as the design function, to a low cost foreign location.

Procurement as a function will, in many instances, use agents or other organizations to fulfil procurement activities. They will do so when it is economic and when they do not have the appropriate personnel available in the right locations. However, outsourcing should not be regarded as a separate process but as an integral part of the procurement function.

The setting up of the arrangements for the specialisms described will involve a procurement process. However, their benefits and the problems in implementing them are outside the scope of this book. Suffice it to say that outsourcing the whole procurement process is a route to losing competitive advantage and discovering, at some stage, that you wished that you had such a capability.

THE BUYER

The person or individual performing the purchasing function is usually called a Buyer and will work in a procurement department or a group within a project team. On smaller projects the buyer may also be the procurement supervisor or coordinator.

Whilst a buyer manages the materials and equipment orders, they can also be responsible for the contract for the services provided by vendor servicemen, who will be commissioning their own equipment on site. The buyer then manages the contract in conjunction with the purchase order.

Whilst many project personnel could perform the middle part of the purchasing process – the more procedural aspects of inviting tenders for goods and services – the buyer brings two key areas of expertise. At the front end of the process they are the source of market knowledge and have the relationships with the suppliers. At the back end of the process they are the experienced negotiators finalizing commercial deals. Whilst all project personnel negotiate to some degree or other, they are rarely trained in negotiation skills. The buyer, on the other hand, practices the skill on a regular basis.

A principle of project management is that there should be one person in charge; namely, a project manager. A primary reason for this is to control communications across the contract interface with the client. Similarly, the buyer must be the person responsible for the contract interface with suppliers.

A primary duty of a buyer is to develop and maintain relationships with their suppliers through the preservation of high standards of professional conduct in all of their dealings with them.

The buyer must make sure that specifications, drawings and information are available for transmission to suppliers at the right time. More than this, they must check that the requisitions describe accurately and unambiguously what is wanted in terms of technical and delivery content.

In the following quotation[2] the term client is used to refer to an individual or user department in the buyer's own organization – although it could also be used to refer to the external client.

...a conflict facing the buyer grows out of the division between the buyer's abilities and his clients' wishes. Given a limited amount of time and money, the buyer can only purchase what the market place is capable of supplying. The precise total specification of the product or service that the client would like the buyer to procure (including price, quality and delivery) is frequently unobtainable. Tapes of the Old Rolling Stones song, 'You can't always get what you want' should be made available for all buyers to play to unhappy clients while their calls are put on 'hold'!

These fundamental facts about the buyer's business life reveal his two basic activities:

- *Persuading suppliers to do things they do not want to do, for nothing.*

- *Persuading clients to accept less than they have asked for, without feeling let down.*

These quintessential activities reveal, in their turn, the buyer's most important skill – he must be a good persuader. The degree of skill actually required in the performance of the buyer's task is determined by the nature of each individual transaction he has to undertake in the two areas of activity – buying from suppliers and servicing clients. The difficulty of each transaction varies enormously.

TERMINOLOGY

Some emphasis has been placed, within this book, on the meaning of words. Consequently, it is necessary to explain how terminology that is in common usage in many different business contexts has been used.

---

2   'Dealing with powerful suppliers' by John Ramsay. *Purchasing and Supply Management*, January 1987.

Whilst the terms Bid, Bidding and Bidder are used extensively, they have been avoided, as far as possible, since their more strict meaning relates to a different process to that with which we are concerned. Bidding is a term used more for auctions in the broadest sense, for example, conventional auction rooms or buying goods on the Internet. In the press and financial world the words are used for the purchase of companies, shares and assets. The terms that apply during the enquiry stage, for the placing of orders or contracts, are: Tender, Tendering and the more cumbersome Tenderer.

It is generally accepted that orders do not involve site or field labour. Any activity that involves the provision of labour, at a location or on a site, is a contract. Thus, the term purchase orders or orders will be used to identify the contract for goods not involving any services using a company's standard contract terms. The term Contract or Subcontract (when placed by a contractor) will be used to identify the contract for services using either standard or tailored contract terms.

The two basic contract types

Contract categories and specific contract types are covered in detail in Chapter 4, however it is useful to define briefly the two basic types at this stage. At one extreme the client accepts the risks by reimbursing the cost of the service provided by the supplier or contractor. A Client Managed Risk Contract. At the other extreme, the supplier or contractor supposedly accepts the risks involved by delivering an end result for a fixed price. A Contractor Managed Risk Contract. In a BBC[3] Panorama programme, entitled 'Bad Deal for Britain' analyzing the Nimrod Airborne Early Warning System project, the reporter and presenter Tom Mangold explained these contracts as follows:

> *The Ministry of Defence Procurement Executive (MOD-PE) had the power to buy Nimrod for the RAF on one of two contract systems. Either a fixed price contract (which means the full contract, in terms of specification and price, are agreed in advance) or a cost plus contract (which means that month by month the firm charge the Ministry whatever costs they incur and then charge an agreed profit on top of that). MOD-PE chose to buy Nimrod on a cost plus contract.*

The words Fixed Price, used in relation to contracts, means that the price does not change for a defined scope. This meaning has been chosen in order to cover a wider section of business and industry contexts. In the process industries, (represented by the Institution of Chemical Engineers standard contracts) the term Lump Sum is used to describe the same contract type. The more generic term, Reimbursable Cost, will also be used instead of Cost Plus.

The Nimrod project will be revisited later to illustrate how there was, and still is, a lack of understanding regarding the contracting process. However it should be stated that, despite the project title of Nimrod, there was nothing wrong with the aircraft. It was the computers that did not work.

Other terminology

The term Client will be used to identify the party initiating the purchasing and contracting arrangements. This can be both the project owner of a facility or a contractor initiating purchasing activities on behalf of the owner. For example, the project owner acts as the client when they purchase goods and services, but a contractor can also be a client when they employ a manufacturer. Similarly, the manufacturer is a client when they buy components from a subsupplier. In government the client can termed the Sponsor. This term can also be used for the client of a project that is internal to an organization. The term Customer will be used for people, or other stakeholders, external to a project. The term user is, more often than not, utilized to identify the operators of a facility.

---

3   Education and Training Video by BBC Enterprises Ltd. 1985

However, the term will also be utilized to indicate the next party in the work process chain who uses the work produced from the previous function.

The terms functional manager and line manager are one and the same. It is generally recognized that the term, ,manager of, say, procurement is a functional manager with line management responsibilities for that particular discipline. Whereas, it is custom and practice in certain industry contexts to use the term procurement manager to distinguish the role of the person responsible for procurement on a project. They may have line management responsibilities for the people within their project team but only for the duration of the project. They will report to the project manager and also to the manager of procurement, that is, they work in a matrix organization (discussed later in Chapter 2).

The words Buyer and Purchaser are synonymous, as are Seller, Supplier and Vendor. The words buyer and seller tend to be the words used in contracts by the legal profession. The term buyer can be an individual as well as an organization, whereas the term purchaser tends to be used to describe an organization. Whilst the term vendor is commonly used to describe the organization supplying the goods (after order placement) it is also used in conjunction with other words to describe some of the activities performed by them, for example, vendor servicemen and vendor drawings.

Whilst the term supplier can be used generically to describe the supply of goods and/or services, it will tend to be used to indicate the supply of materials and equipment. Similarly, the term Manufacturer will be used for fabricated items, for example, rotating machinery, electrical cabinets or window units.

Engineered or designed items are items of a unique nature that have been designed or specified to meet the particular requirements of a project.

Bulk materials are commodity items that have a uniform catalogue type description or standard material specification, such as: piping, cabling, structural and civil building materials.

Services are provided by consultants (for example, architectural, engineering or other specialists) or contractors. Services also include the use of vendors' commissioning and start-up specialists for equipment supplied by them that involves site work. A distinction is made in the project world between those that provide only design type services – consultants; and those who perform a whole range of services (for example, design, procurement and installation or construction) – contractors. They are also distinguished by the risks that they are willing to carry. Consultants' organizations have, in the past, been partnerships and accepted personal liabilities, but have not been willing to accept construction risks. Contractors, on the other hand, tend not to take personal liability but are distinguished by taking on construction risks.

Where imperial units of measurement are used they relate specifically to the process and oil industries.

2 Purchasing and Project Management

THE PURCHASING AND CONTRACTING CONTEXT

The purpose of this chapter is to set purchasing into a project and project management context, identify the processes involved and provide linkages to later chapters.

In non-technical environments, or in organizations with a weak project management culture, it is not uncommon for purchasing to be seen as a separate central services activity. This can be particularly true of organizations whose origins were in Government. It is my view that (apart from the regular change of personnel responsible) the attitude produced by the focus on procurement, of say defence type projects, is one of the key reasons for their problems. The Government believes that it is procuring a product when, in fact, they are the client of a project and need to project manage its development.

One can purchase materials – office supplies, sand and cement or piping and electrical fittings. One can even purchase equipment – photocopiers, vehicles or pumps and compressors. Despite the statement that one can purchase these goods they must still be specified (see Chapter 7, Communicating the Requirements). However, one does not just purchase services. Design and construction of a new building, engineering and construction of a new plant or installation of a new system all involve buying services for their development. They all require more than purchasing; they need to be project managed.

In purchasing goods and/or services in an organization we will have a purpose in mind, and that purpose will involve a project. In addition, each purchasing activity is a project in its own right and the buyer is managing a portfolio of projects. Consequently, we need to understand the process involved in developing a project and when it is most appropriate to perform the purchasing activity.

In order to control the project management process of developing a project, it is broken down into stages or phases. The number of phases will determine the level of detail at which control is exercised. Between each phase there is an opportunity to assess the viability of the project and decide if one wishes to proceed to the next phase. The same is true of the steps in the purchasing process.

Project phases

Typical project phases are briefly explained below and are illustrated in Figure 2.1 together with the principle management activities carried out at each stage.

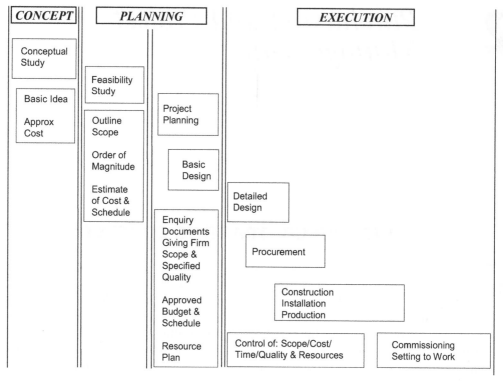

Figure 2.1   Project phases

Terminology in a technology environment has been chosen in order to cover as wide a context as possible. Different industries and business contexts use different terminology, and the length or duration of the project phases vary – the words, phases and stages are one and the same.

These are the fundamental phases that any project must go through to be successful. Leaving one out, for example the feasibility phase, is a recipe for disaster.

Concept: a company, government or some other body determines that there is a requirement for a new facility, plant or product.

Feasibility: the concept is examined in detail to see whether it is a realistic, viable proposition.

Planning: if it is viable, an execution plan is developed.

Basic design: before major funds are committed, the basis of design is carefully agreed.

Design: once the basic 'recipe' is firm, detailed drawings for each component are produced.

Procurement: all the necessary materials, equipment and services are bought.

Fabrication/
Construction: the facility, plant or product is assembled from the materials and equipment using the drawings prepared.

Commissioning: the facility, plant or product is thoroughly tested to ensure that it satisfies the requirements of the project and is set to work.

Implementation or Execution are terms that are used to describe all the phases of a project after it has received approval to go ahead, or after a contract has been awarded by an owner client to a main contractor. Implementation, however, for Information Technology (IT) projects is used to describe the transfer (including training) of a software system into an organization. In process plant projects this is the commissioning or the setting to work phase.

The process of starting a project is generally termed Project Launch. This is the period of time where the project personnel go through a learning curve process, get organized, establish procedures and gather information. The end of the launch period is when production work has been established with a consistent rate of progress. Project launch can be applied to the start of any of the main phases. However, in product development the launch phase is at the end, when the product is launched into the market place. In effect, the equivalent of the IT implementation phase. Some people use Start-up for the launch process. However, start up in the power and process industries is at the end – the commissioning or setting to work phase.

Sometimes the phases are carried out sequentially but very often, because the project is required more quickly, the phases are overlapped.

It can be seen that the purchasing process is part of a phase in its own right – the procurement phase. Obviously, this activity cannot take place until the material and equipment has been specified and designed in the previous phases.

One can, however, purchase goods and services at any stage. If they are purchased earlier than the logical phasing, then there is a risk involved. This may, however, be necessary for items that have a long manufacturing and delivery period. If they are purchased later there will be excessive expediting costs and there is a risk that the project will be delayed.

Purchasing can be perceived as an office-based activity but most projects involve activities in more than one location, particularly technological type projects. Only 20 per cent (although 80 per cent by value) of the requirements for an engineering and construction project will be purchased in the home office. 80 per cent of the purchasing (20 per cent by value) will be carried out at a remote location or on a site. Because of the importance and high value of the materials, the personnel in the home office, who are remote from the physical activities of the project, often start to believe that they are producing an end product in its own right. This is, of course, not the case – the home office is providing a service to the later phases of the project. It is the production, manufacturing, fabrication or construction people who are actually producing the end product. Further, as the people who have to interpret the design, they need to be involved in the earlier definition stages. Additionally, the people in the home office need to constantly bear in mind the fact that small mistakes in the early paper development phases can cause a vast amount of work in the later physical execution stages.

In the past, owners would perform much more of the work involved in developing a project and might perform the majority of the purchasing of the equipment for a project. However, since organizations have shrunk back to their core business, owners have passed more of the responsibility for procurement on to contractors.

Should the owner wish to there is an opportunity, between each phase, to invite contractors to provide the necessary design and construction services for the development of the subsequent phases. Traditionally, the enquiry process for appointing a contractor has taken place after the feasibility, planning or basic design phases. However, in engineering construction process-type projects, the stages in the procurement process for these services are not regarded as project phases in their own right.

Owing to the lack of expertise of owners in building work, it is necessary to pass the management process over to a third party. Consequently, the owner employs an architect to interpret their requirements and represent their interests during the development of the project. How well, or poorly, this is done is not a subject for this book. The Royal Institute of British Architects (RIBA) has what they call Plan of Work Stages A to L. In this context it is essential to go through a process for awarding contracts for the necessary construction services. Consequently, the invitation and evaluation of tenders and the contract award are treated as stages in their own right.

The RIBA stages are roughly equivalent to the more conventional phases as follows:

Stages A, B and C: concept and feasibility.
Stages D and E: project planning and basic design.
Stage F: detailed design.
Stages G and H: inviting and evaluating tenders.
Stage J : letting building contract.
Stages K and L: construction and completion.

Purchasing cycle

The purchasing process is identical whether from an owner's or a contractor's perspective, since when a contractor purchases goods and services they are acting as a client.

The following Figure 2.2 shows a conventional purchasing cycle during the procurement phase. To enable the purchasing group to identify and contact suitable vendors, the design department should advise the purchasing group in advance of approximate quantities of materials and types of equipment. Whilst the design department is finalizing the requirements and preparing the requisition documents (see Chapter 7, Communicating the Requirements) the purchasing group is checking whether the vendor has the requisite capability and is likely to submit a competitive price and an acceptable delivery. This process is called Pre-qualification (see Chapter 8, Selecting the Tenderers). At the same time as this pre-qualification process is being carried out the enquiry documents will be prepared. Once the tenders are received copies are given to the design department for a technical evaluation, and purchasing carry out a commercial review (see Chapter 11, Evaluating the Tenders).

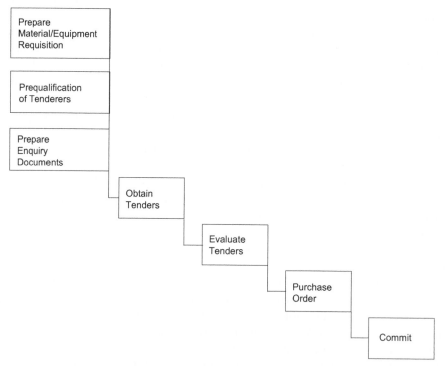

Figure 2.2   **Typical purchasing cycle**

There are a wide number of variations to the purchasing cycle. The next two figures show two extremes. The process shown in Figure 2.3 might be used in order to save time. In this situation purchasing selects a vendor with the requisite manufacturing capacity and negotiates a price for the goods in order to eliminate the tendering and evaluation periods. Offset against this is the time spent negotiating and the possibility that the price may not be as keen as if competitive tenders had been obtained.

The opposite extreme in Figure 2.4 might be used when the project owner is using a contractor and wishes to retain control over every aspect and be able to justify, to external auditors, that the correct procedures were used. This formalized procedure is obviously very time consuming since the contractor has to obtain the owner's approval at every stage. It was used extensively in the Middle East at a time when there was a lack of trust between the owner and contractor.

The steps involved in the contracting cycle are very similar to purchasing equipment and materials. However, it is necessary to highlight the importance of deciding the allocation of risk and, hence, the appropriate contract strategy (see Chapter 3, Contract Strategies).

In identifying the enquiry process for materials and equipment we have only considered the client's or purchaser's perspective. It is important to recognize that two parties are involved, and the following illustrates the purchaser's activities and the tenderer's responses to them:

INITIATOR	**TENDERER**
Client/Purchaser	Contractor/Supplier

- Project Initiation/Definition   Marketing
- Decide Contract Strategy
    - Allocation of Risk
- Select Tenderers   Pre-qualification
- Issue Enquiry   Tender or Proposal
- Evaluate Tenders   Presentation

                   • Negotiate Contract

                   • Place Contract

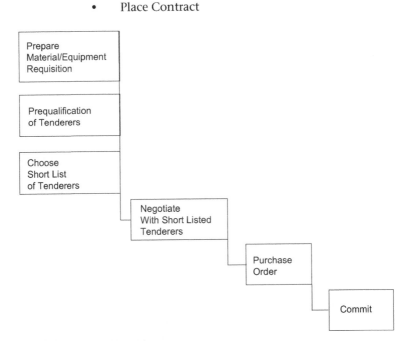

Figure 2.3   **Negotiated purchasing cycle**

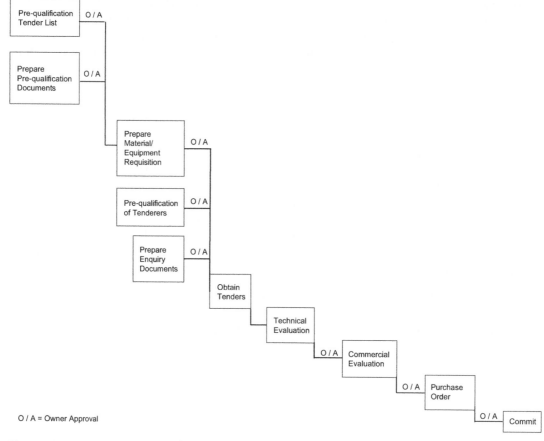

Figure 2.4   Owner approval purchasing cycle

This is very much a simplified process only showing the main activities. In reality there are many more detailed steps of a more administrative nature, and these are dealt with in Chapter 9, The Enquiry Process.

Involvement of users

A fundamental criterion for success in project management is that the Users must be involved in the early project development phases. The term user defined earlier is the next person in the chain of the work process. The user of the feasibility study will be the project execution team. Similarly, the user of the design will be the fabricators or manufacturers. The subsequent users will be the construction and commissioning teams and, ultimately, the owner's operators.

One of the flaws in a supplier or contractor's experience is that they are not involved in operating their products. It is for this reason that owners of process type projects want to have their operators involved in the design stages – with consequent implications on the type of contract. Manufacturers of equipment are more effective, in this respect, than design contractors since they do at least service their equipment and, hence, gain experience with operating problems. The design and construction contractors are next since they should, at least, be able to get the design to suit their own construction people. The lone designer is worst of all since they not only lack operating experience but they also lack construction or installation experience – the classic problem with the architect designer.

One contractor's organization I know well, M W Kellogg, takes the entirely sensible approach of saying that designers and engineers should be responsible for buying the equipment that they have requisitioned. They then follow their orders through on to site and become the site engineer responsible for installation and start-up. This addresses the issues of eliminating interface problems, involving users, and providing designers with commercial, installation and commissioning experience. The main problem is finding people, who can be rooted psychologically in a home office culture, willing to go to a remote location.

This requirement to involve the users poses a dilemma since, in the early stages, the main contractor or key suppliers may not have been selected. Consequently, a different contracting arrangement, to the traditional confrontational competitive tendering, is required. (See Chapter 5, More About Contracting).

Organization options

Organizations have developed into functional groups or departments in order to perfect work routines and increase the certainty of producing consistent and high quality deliverables. However, the organization will have a mission statement defining its overall purpose, its scope and its boundaries. In order to achieve this mission statement it will be necessary to take specific actions – deliver different and more valued products, enhance the organization's skills with training, provide specialist consultancy assignments, as well as introducing new systems. Unfortunately, none of the projects resulting from these actions fit within the functional departments. This is because the projects need to be integrated across all the departments in order to achieve the project objectives. There is now a dilemma. Either the business concerned has to change its organization structure, or change the way it works. Should the organization face the outside world with an image of 'look how well we run departments' or should it rotate through 90° and face the outside world saying, 'Look how good we are at managing projects'? In either case, or some compromise in between, projects and departments create a matrix organization and *it cannot be avoided*. The type of project will influence the choice between department emphasis and project focus.

In purchasing goods and services two managers are involved; the functional manager of procurement or purchasing and the project manager. Consequently, we have a quandary as to how the procurement work should be organized. Should it be centralized in the functional department, should it be located in the project or should it be some combination of the two? There is no right or wrong answer. The issue will be debated by the project team and will depend on issues such as project size, duration, complexity and uncertainty. Project managers will have, or should have, a preference for some procurement presence within a co-located project team. For those readers wishing to go into greater detail concerning matrix organizations the following reference is recommended: Organizational Alternatives by Robert Youker – Presented at the 8th Annual Symposium, Project Management Institute Montreal, Canada, October 6–8, 1976. This seminal paper forms the basis of all that has been written on the subject since that date.

Once project launch is complete and production work is well established (but before it is too late) the sensible project manager will use the matrix, and request one or two functional or line managers to carry out audits of the work performed. The audits will be phased or scheduled at different times and the work is often delegated to senior, experienced personnel working on different projects to the one being audited. They will check the quality of the work and check that corporate and project procedures are being adhered to.

As previously mentioned, departments develop routines and there is no precise boundary between routine management and project management. Consequently, as the work moves away from being routine and becomes more project orientated, it may still be appropriate to manage the project in a functional department, that is, a functional matrix. Small projects that are very similar to departmental work are called Runners. Projects that are of a similar or repeat nature are, naturally, called Repeaters. With repeaters the work has become more specialized and less routine – more project focused. Consequently, a balanced matrix is likely to be more suitable. There is a balance between the departments taking the lead but coordinated by a project manager, and the project manager taking the lead but still supported by the functions. As the projects become more unusual they become Strangers to the organization's normal method of working, and a dedicated project team (within the matrix) will be the preferred option. The secret to success of an organization is to break the project down and turn the processes into less complex types. That is, turn strangers into repeaters, turn repeaters into runners and runners into Routines. This method of categorizing projects will also influence how we organize the procurement function.

Manufacturers and suppliers tend to be specialists in their own particular fields and organize themselves accordingly. Consequently, it is logical for purchasing organizations to mirror these specializations and organize themselves into commodity groups. In effect, turning runners into routines. Each company will have its own variants of these groups. For example (in the list below) instruments and electrics may be divided into two separate groups. Typical groupings are:

- mechanical/rotating equipment;
- heating, ventilating and air conditioning equipment;
- instrument and electrical equipment, including bulks;
- pipework and valves;
- structural materials;
- fabrications;
- packages and special equipment;
- architectural materials;
- and so on.

A chief buyer will head each commodity group. They will be experienced in the capability, performance and reliability of the suppliers in their area of responsibility. They will follow price trends, delivery lead times and manufacturers' workloads and capacity.

Since each purchasing activity can be regarded as a project in its own right, we also have to decide how to organize the matrix within the procurement function. The purchase of commodity-type materials can be perfected into a routine, and the purchasing projects can be treated as runners. Thus, the buying of materials may be best carried out in a functional group.

Buying equipment, on the other hand, is more specialized but is still, however, of a repeat nature – thus one of the balanced matrix options, with more project involvement, is more likely to be preferred. However, in this situation it is also necessary to prioritize the programme of work (the portfolio of small projects) for the allocation of the scarce resources at the disposal of the buyer. In purchasing, the priorities are assigned by the demands of the schedule and anticipated delivery periods.

The purchasing of services is much more complex – every contract is unique and non-repetitive so that the contracts buyer is, in effect, the project manager of the subcontract. Consequently, these strangers are best done by a small team of dedicated people and a local procurement group may formulate the contracts that require installation or construction work.

Planning decisions

It would be wrong not to mention that before starting the purchasing process it needs to be planned. The project procurement manager will be working closely with the project controls personnel. They will be validating the estimates and breaking down the summary schedule into detailed timescales for the purchasing activities for the equipment, materials and contracts. The following questions will need to be resolved:

- How long should be allowed for each stage in the purchasing cycle?
- In particular, how long should the tendering period take?
- What are the budgets for all of the goods and services to be purchased?
- Who is responsible for originating the documentation required for each activity?

In addition to the above issues, the tender list will need to be validated as well as a number of questions concerning the competence of the suppliers or contractors. These issues will be dealt with in Chapter 8, Selecting the Tenderers.

Further, there are also a number of strategic options (discussed in later chapters) that will need to be resolved in conjunction with the project manager and the rest of the project team:

- the contract strategy to be used;
- the contract terms to be used;
- the tendering method to be adopted;
- the basis of the evaluation criteria.

RISK, RULES AND RELATIONSHIPS

It is important to be clear about the reasons for purchasing goods and services. The primary reason should always be that someone else is better at performing the required process than we are. This is fairly obvious for the supply of manufactured items such as materials and equipment. We may not have the facilities. However, for the supply of services the answer to the problem may be questionable. Using someone else to provide a competence that we do not have obviously involves taking a risk.

All contracting arrangements involve some degree of *risk*. Due to the different contractual and commercial arrangements between the parties there will be a need to write different terms for the contract – *rules* that will determine what happens when something goes wrong. From this it clearly follows that a *relationship* will exist between the parties. I have borrowed the concept of these three words from my friend and colleague David Wright but I like to illustrate them diagrammatically – Figure 2.5. A diagram can show the interaction between them and show how they change under different circumstances.

The size of the circles can be used to show how each element impacts on each other. The greater the risk the more precise the rules will need to be about the division of the risk. The greater the risk

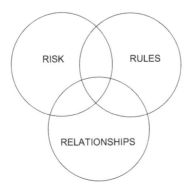

Figure 2.5   **Risk, rules and relationships**

that is transferred then the more distant the relationship will be. The more risk that the client retains then the more integrated the relationship will be.

The three elements need to be in balance: if one is overemphasized at the expense of another then the contract is likely to have problems. The risks and rules have always been clear and dominant. It is the relationships that have been ignored or suffocated in the short sighted competitive environment. As Napoleon said, 'War is one-quarter logistics and three-quarters relationships.' It is possible to have a perfect set of rules, but with poor relationships the contract or project will fail. On the other hand, with an onerous set of rules and good relationships one can succeed.

Risk

Risk comes in many forms, for example, technical, commercial, physical, environmental and so on. The Association for Project Management (APM) Body of Knowledge[1] defines risk as, 'An uncertain event or set of circumstances that, should it occur, will have an effect on achievement of one or more project objectives.' However, it is not intended to discuss the subject in any detail here. Risks should be borne by the party best able to:

- estimate the risk;
- manage and control the risk;
- carry the risk.

The risk should not be too great for one party or outside the control of either party. In order to obtain a satisfactory level of competition in the contracting process, the risks should be modified to the extent that the contractor can carry or manage the risk. Risk is a function of the unknown elements of a project and is directly related to scope definition. Consequently, as size and complexity increase there is a tendency to move to a form of contract that provides a more equitable allocation of the risk. The allocation of risk in contracts and who is best able to manage and influence the risk will be analyzed by the core project management team, and is discussed in Chapter 4, Contract Categories. Nevertheless, a buyer should be wary of anything with one of the following characteristics:

- new or novel;
- innovative;
- the first of a kind or never done before;
- technical uncertainty;
- the biggest, a large scale-up or the unfamiliar.

If they spot a requisition with any of these elements, or that might exist or be generated for the supplier, they should check what special action might be required in the purchasing process.

A typical risk that a project manager might initiate is pre-empting the normal purchasing processes for significant cost or schedule advantage. For example, we had just signed a Fixed Price Contract in our client's offices in South Wales and I remember stopping at the first available opportunity to ring our office. As project manager, I instructed them to place an order, the same day, for some high-value lead-lined electrostatic precipitators. The order was to be based on the quotation that we had received during the tendering period and that we had used in our own tender. Being confident of the capacity, no further validating design work was to be done – we would develop the rest of our design around what we ordered. We took a calculated risk and it worked. It made the difference between a project that would run late, with consequent cost overruns, and one that was a successfully completed on time.

---

1   5th edition. See *Additional Sources and Contact Details*.

Similarly, at the height of the North Sea developments, it was not uncommon for owners to place orders for the jacket steelwork, in order to reserve mill capacity, before they had awarded a design contract.

The rules – the contract terms

There is a direct connection between the definition of the scope of work and services, namely the task to be performed and the involvement of the client. If the task is clearly defined then a more hands-off approach can be taken to the contract. With less definition the client needs to be more involved until the definition is so weak that the owner needs to take control. This relationship between the parties needs to be reflected in the words of the contract – the terms.

It is the people who define the rules of the contract. They do this when they have a common objective. They define and agree the rules that should be used for when something goes wrong.

Nevertheless, one should not put oneself at a disadvantage by negotiating with an essential partner after a higher level or main contract has been signed. For instance, we had tendered for a plant on the basis of using two proposed partners. One partner was locally based (an important criteria) and the other was essential since they were one of the few contractors with some vital proprietary technology. The deal was set up on the basis of a shake of a hand! If we won the contract they would accept the same contractual terms as we were able to negotiate with our client. The combination was a winner and we signed a contract with the plant owner. The locally-based contractor fulfilled the spirit of and the letter of the handshake, whereas the specialist contractor imported two American lawyers. What happened next should never have been allowed. As a company, we did not have a lawyer on the staff but we did have someone who specialized in contracts. However, for reasons I do not recall, he was not available to help. Consequently, as project manager, I was given the task of negotiating, on my own, with the two lawyers. The result was, that for a week I was, what I can only describe as, legally raped! I still blush at the memory of my naivety. First lawyer, 'We really don't like the wording here.' My response, 'Well, that's unlikely to happen.' Second lawyer, 'If it doesn't really apply, it should be taken out,' and so on! In my defence I did try to limit the damage, and on checking with my client I was told that we had only been awarded the contract on the basis that we used this particular subcontractor. I then explained what was happening to the general manager and suggested that we pull out of our contract, as we were being forced to accept unreasonable risks. His response was, 'Do your best.' I did, but it was a poor best at that time. I did not have the negotiating skills, I did not know enough about contracts and I did not know enough about the law. Fortunately, everything worked out well.

The project manager must know enough about the rules of the contract to be able to talk effectively to a corporate lawyer. If a subcontractor is an essential part of a proposal the key contract terms must be negotiated before a tender is submitted. Obviously, individuals should not negotiate contracts on their own, and I would even suggest that project managers should not actively take part in the contract negotiations. They should certainly be consulted and make suggestions, even direct the negotiations from a back room. My reasoning is that it might be difficult to maintain relationships during the operation of a contract if the project manager has just been a hard-nosed so and so. It is easier to blame the lawyers and for the respective project managers to stay friends and say, 'Let's try and make it work for us.'

Relationships and the marketplace

One of the buyer's duties is to get to know the suppliers in the marketplace and determine which ones are suitable suppliers to have on an approved list.

At some time or other we have all had a bad experience with a supplier. If we have felt sufficiently strongly about it we have reported back to the procurement department saying, 'Cross them off the list – we should never use them again.' Some period of time later, to our surprise and horror, one of our colleagues is using them. Why is this? This is so important that we need to understand the issues involved. The reason for using the same supplier again can be justified in a number of ways:

- we have got to know them;
- we understand how they work;
- they are familiar with our procedures;
- we know where their weaknesses are and where they need additional support;
- we have been through the learning curve;
- better the devil you know;
- and so on.

This implies that the overriding issue is the *relationship*. Even a bad relationship is better than no relationship. 'We have learnt from our mistakes,' they tell us. This is important to the buying organization; organizations do not want to go through the same learning curve all over again. However, there is a fallacy to this rationalization. We do not buy from companies. We buy from the people in the company. So that the justification arguments are only valid if we are dealing with the same people, and on most occasions we are not. When buying standard materials or equipment it is less important and there is a chance that the supplier allocates our order to the same individual in their organization. If we are buying services the chances of having the same team of people are much less likely, unless as we should be, we are dealing with a smart contractor that understands these issues.

The result of recognizing the importance of the relationship is that we will want to get to know the suppliers in some depth. We will also expect to get improved service, preferential treatment and a reduction in the risks involved. However, the supplier will be reluctant to go down this route if they are one of many suppliers. They will expect to be part of a select few. They will want to know that the extra effort invested in the relationship will result in more orders. The consequence is that, if we want to improve reliability of supply and we want enhanced performance of services, we need to build and enhance the relationship and 'get closer to fewer'. Unfortunately this option is not available to the public sector, which may be a contribution to their poor performance.

During the procurement phase of the project there will be a closer relationship between the project manager and the individuals of the procurement function. The project manager will be more involved and more familiar with the details of their work, than with the other disciplines. So that, as well as external relationships, it is important to maintain internal relationships, and hence communications, throughout the team structure. This needs to be done without upsetting the supervising hierarchy.

The following example illustrates how internal relationships are as important as external ones. A field expeditor and inspector telephoned me saying, 'I am ringing you direct because I think this needs your attention. I think you should come and look and assess the situation yourself.' My first reaction was of a negative nature (why were they not reporting through the normal channels?) but then I thought this person would not ring me direct unless it was important. Within a few days, when I arrived in Holland, the vendor I was visiting was in administration (similar to Chapter 11 in the USA). Seeing our water-tube boiler in a thousand pieces spread across the factory floor I realized that our project was dead unless we did something to keep the company going. Understanding that the Administrator needed cash, I arranged to buy the materials for cash and put the owner's name on them – we could then take them away if we had to, rather than having them sold off by the Administrator. New payment terms were negotiated together with a bonus for early completion. The vendor survived. Our boiler was delivered early resulting in an early project completion date and a bonus from our client.

The project management role

Perhaps the most important relationship is the one between the client and their project manager. For this reason the sophisticated owner client will relocate to their main contractor's offices. However, there comes a time in every project when the centre of activities moves from the contractor's home office to the owner's location or site. Since the contractor project manager should be where the client is, this is when their project manager moves to the remote location. Once the focus of activities has moved to the site most of the purchasing is more likely to be of a local nature.

Every country in the world has some local capability or other. In particular, they all have a capability in building and civil engineering – whether good, bad or indifferent. Less developed countries will be particularly keen for this expertise to be used. Principally, because it reduces the requirement for imports, which would use up valuable foreign exchange, but it also generates employment and enhances local capabilities.

It may be perceived that the project manager is at the peak of a hierarchical organization structure in charge of all that they survey. However, this is not correct; the project manager's role is a supporting role. They support their key managers who are responsible for making sure that the work of the functions is performed correctly and on time. I will argue strongly that the project manager should not get involved in areas that are the responsibility of specialists. In reality, matters often go awry in the project manager's area of expertise because they interfere – thinking they know something about the subject! The project manager's role is to make sure that the people doing the work have all the necessary information, materials and facilities they need. Consequently, they will have a close relationship with the project procurement manager, monitoring the release of information from vendors, resolving problems and making sure deliveries will happen on time. However, there is always something crucial, different or special on all projects. As a consequence, the project manager can be more intimately involved with the purchasing of some key items, and the departments involved may resent this.

The following experience illustrates the complexity of procurement issues that the project manager might be involved in, and where the project manager *can* be justified in taking command and directing the course of events.

On a project that I was managing, I knew that project acceptance was at mechanical completion. I also understood the conventional meaning of the term as: some combination of testing everything and getting it ready for operation but without using process fluids. Consequently, as we were approaching the end of the project I decided that it was about time to read the contract. I needed to be thoroughly familiar with what was necessary to achieve mechanical completion and, hence, project acceptance. Imagine my surprise, shock and concern to realize that in order to achieve the contractual description of mechanical completion, the plant had to be fully operating, making product. Surprise at the contractual description of the term mechanical completion shock that I had not looked at this aspect of the contract in detail before – I believe in being familiar with the contract but getting on with the job and not referring to it on every occasion. Concern that there was sufficient time to do whatever was necessary.

Fortunately, our project commissioning manager had just arrived and I cross-examined him on how the plant was started up. 'You need ammonia,' he said. 'How much?' I asked. He carried out some quick back of the envelope calculations, made allowance for it not working the first time the button was pressed, and came up with the alarming answer of 2000 tons.

I rapidly got in touch with the home office in London to ask where they thought I could purchase 2000 tons of ammonia. I received the answer (which is engraved upon my mind), 'You can probably get some from Madras Fertilisers [in Southern India] in 45 gallon drums.' Now I had long since left any technological capability behind me; but without talking to anyone, I knew this answer, using 45 gallon drums, was useless, impractical and unhelpful.

I was bemoaning my problem over a glass of beer in the local swimming club and the person I was talking to told me that he knew just the right person to help me, in Singapore. On telephoning this contact they would not tell me the solution to my problems until we could work out how he would fit into the deal. I needed him. Not only did he know where to get the ammonia in bulk, but he also knew the key to the problem; how to transport it. He knew where there were enough containers to ship the liquid ammonia under pressure. It worked out that we needed 38 pressure vessels, 30ft long by 9ft diameter! So off I went to try and negotiate something. We agreed that he would supply the charter vessel that was required and this satisfied him. I then had to persuade my client's technical advisor (it was their money we would be spending) and we were off to Indonesia to make our purchase from the national fertilizer corporation. The negotiation took place with an army general using a white board starting with the market price and adding on various costs. I was reviewing the price make-up when I noticed that a 5 per cent had crept in and so got up and wiped it off, as it was not something we had discussed. Ten minutes or so later the 5 per cent had materialized once more, without any comment, and so I removed it again. When it reappeared for the third time I realized that perhaps this was someone's commission! Having agreed the price, including the hire of the pressure vessels, the general decided that he wanted them returned within a relatively short timeframe and demanded a hefty penalty if they were late. That evening, in the Jakarta Hilton, I was writing out the terms of a letter of credit for $400 000.00 (in 1980 that was a lot of money) when it occurred to me that I might not know what I was doing! Consequently, I telexed London and asked procurement to check over the letter of credit in case I had missed something or made a mistake. The response I received demonstrated the sensitivity of inter-departmental relationships. 'What are you doing in Jakarta without a representative from procurement?' The question remained unanswered! There was only one thing for it. Go ahead with the deal, charter the necessary vessel and ship the ammonia. At the appropriate stage I had informed site of the arrangements and the site traffic personnel did a fantastic job of clearing hundreds of obstructions (power and telephone cables, road signs and so on) over 12 miles of road. The ship arrived on time, the containers were offloaded and transported in a faultless operation and the ammonia was transferred to a storage tank on site. The containers were then returned on the same ship. It still surprises me to this day that it all went without a hitch – until the start button was pressed. The machinery of the plant turned over, struggled, coughed and spluttered and stopped. With slight mental panic I asked the commissioning manager, 'What happens if the plant doesn't start up next time?' His answer was predictable, 'You buy some more ammonia.' Fortunately, though, it started and the rest is a different story.

3 *Contracting Strategies*

INFLUENCING FACTORS

As already stated, the purchase of goods and services involves a contracting process. This may seem obvious but the basis of successful purchasing is (after complete specification of the scope) choosing the right contract strategy – the contracting means to achieving the project objective. It is finding the strategy to align the purchaser's objective with the supplier's or contractor's objective that is the key to successful purchasing. It is fairly clear-cut with the purchase of materials – one wants to buy and one wants to sell and it is a matter of negotiating the price. Once we move away from the more straightforward issue of price and start talking about delivery the buyer's and seller's objectives diverge. The seller wants the buyer to collect the goods from the factory and the buyer wants the seller to deliver the goods to the buyer's premises. With equipment it becomes slightly more complicated. Perhaps the buyer would like the seller to include items in the package that are outside the seller's normal activities. In contracting for services it becomes even more difficult to find the contracting mechanism that will align the buyer's and seller's objectives. This divergence of objectives means that the supplier's behaviour is diverted from the intent of the contract.

In order to arrive at the appropriate contracting means, decisions need to be made about the following aspects of a project:

Risk,	which is allocated by;	the basic contract categories.
Scope,	which is covered by;	the extent of execution and the point of delivery.
Money,	which is determined by;	the terms of payment.
Motivation,	which is provided by;	incentive schemes.
and		
Change Control,	which depends upon;	monitoring procedures.

The decision the owner has to make is: which strategy will provide the oversight required to give a safe facility, at the desired level of performance and provide the most effective schedule and budget controls? It is a bit like a game of chess – what opening gambit should be employed? What moves should be made? However, along the way we will win some and lose some but we will always have our long-term objective in mind. Key questions for the client are: How should the risks be divided up? How much risk can be passed on to the supplier? Too often the approach is taken to use a contract form that passes as much risk as possible to the supplier or contractor even though the definition of what is required is not yet finalized. Further, some clients think that they can then change things as they go along. It must be understood that in passing risk one passes control. As a result, the contracting strategy to be used, namely the division of risk and the allocation of risk, must be addressed right at the start of a project. Changes later on will escalate the costs significantly.

Consequently, contract strategy is a process of logical decisions regarding the various options for the:

Division of the Work and the Performance of the Work.

In order to come to a decision on the most suitable options for these components it is necessary to evaluate the objectives, constraints and opportunities that will affect the performance of the design, procurement and execution of the project. It is, therefore, necessary to identify the factors where decisions have to be made, in order to achieve the highest probability of meeting the project, the owner's and the user's objectives.

This immediately poses a dilemma since there is a conflict between the objectives of these stakeholders.

- The sooner a product is in the marketplace the sooner a project generates a revenue stream. The financial modelling of this element is all-powerful. Hence, any project that delivers end products is driven by *Time*.
- The owner is probably worried about the financial viability of the project. Is this project the right project to invest in? Should they invest in a lower cost project? What does the cost-benefit analysis show? Can the benefits be increased and the costs reduced? Hence the owner's objectives are more likely to be focused on *Cost*.
- Not surprisingly the user is interested in having more features, more facilities with better quality equipment and materials in order to reduce maintenance. Hence the user's objectives are focused on *Performance and Quality*.

The above has been oversimplified in order to make the point that, unfortunately, as we know from the classical project management 'Cost, Time, Quality' triangle, we cannot achieve all three at the same time. As stated in the Holyrood Inquiry Report into the cost overrun and the delays in the construction of the Scottish Parliament building, 'The press release announcing the competition referred to an early timeframe, value for money and quality. That was the architectural equivalent of motherhood and apple pie. Who would not want all these architectural and economic virtues?'

The way we propose to organize the contracts – the division of work and the type of contract, namely the performance of work, is influenced by three factors: our own organization – the *Internal Forces*; the nature of the project, that is, its *Characteristics*; and lastly, but by no means least, the *External Forces* acting on the project. Issues affecting these three strategic forces are tabulated in Figure 3.1.

The following provides examples of some of the issues listed in Figure 3.1.

Internal forces

Funding institutions and banks do not want to take risks. Thus a company that has a good reputation for completing and delivering projects on time and within budget can find that work is negotiated with them or 'given' to them. As a result they might not have the resources, experience or expertise for competitive tendering.

Project characteristics

A project may be of such a size that it uses up all of our organization's resources. Consequently, we may want to contract out more of the work. Alternatively, the project size and complexity may be such that it is inappropriate to give the work all to one company. It will be better to break the project up into smaller elements.

Figure 3.1   Strategic forces

If the project is in an early phase there may not be enough information to let a Fixed Price Contract.

If a short schedule is required it might be more effective to start work immediately and develop the project in conjunction with the contractor, using a reimbursable form of contract.

External Forces

A company may be traditionally risk averse and/or have been fortunate to obtain reimbursable work during times when there was a lot of project activity. Alternatively, historical precedent may exist. The UK Government got its fingers burnt with the Nimrod Early Warning System Reimbursable Cost Contract and reacted by saying they would only have fixed price contracts from then on. However, this did not solve their problems since they did not understand the requirements needed to manage fixed price contracts effectively – namely, full scope definition and do not make changes. Difficult with military projects!

Market conditions are one of the bigger forces determining how we decide on the type of contract. For a number of years prior to 2004 it was, in general, a fixed price marketplace. This must be qualified by saying that this was not necessarily true across the full range of projects. It will have varied according to the investment plans of different business or industry sectors. Let us examine what determines the type of marketplace, since it has little to do with the clients' desire for certainty of price. We will start with an organization that is full of work and along comes a client with a project where most, if not all, of the risks are to be the tenderer's responsibility. The response from the tenderer is likely to be a polite 'no thank you'. On the other hand, they could say we might consider your project if you, the client, take the responsibility for the risks – assuming the rates for reimbursing the costs of personnel are comfortable or even generous! Let us now consider the same organization but it has lots of projects just coming to an end. As a result of which they are fairly desperate for new work because there is not much in the marketplace. The same client asks them to tender for the same high-risk project. The response is likely to be one of 'yes please' grabbing the opportunity with both hands. In effect, the organization prostitutes its attitude to risk depending upon the market conditions. If a contractor's organization is full of work they will be more risk averse. If, on the other hand, they need work they will be more willing, or forced, to take on more risk.

DIVISION OF THE WORK

To be successful one must recognize that one should not manage a large project – one should manage smaller projects. How many large public sector information technology projects have been successful? I can name one – the original computerization of Pay As You Earn (PAYE). Consequently, the starting point for developing a contract strategy is the Product Breakdown Structure (also known as PBS). 'The product breakdown structure is a product orientated hierarchical breakdown of the project into its constituent end items or deliverables without the work packaging activities attached. It stops with the product end item definitions.'[1] This concept is developed and defined in more detail in Chapter 7, under 'Scope and Work Definition'.

The decision of how to break a project down into smaller projects is often decided for us by its natural decomposition into major chunks or subsections at Level 2 of the product breakdown structure. These subsections are more often than not by physical location or geographical area. This is just one of the two basic dimensions for dividing up a project. The other principle dimension is by project phase.

As an owner, or contractor acting as a client, we are less likely to break down the supply of equipment into smaller packages due to the risks at the interfaces, and unless the client assumes the risk, there would be no single person responsible. On the other hand, the equipment manufacturer acting as a client will want to break down their project into subprojects, as already described, in order to purchase subcomponents. However, the owner or client will want to identify the material components for a project at the lowest level of the breakdown structure.

The next strategic decision relating to the Division of Work in order to achieve our overall objectives is '*What should be done in-house of what should be let out?*' In effect a make or buy decision. Should the manufacturer make the components for their product or place an order with a subsupplier – as is common in the motor vehicle industry? With a contract for services, design development, basic design, detailed design and execution might all be done by the contractor. Alternatively; design development could be done by the client, basic design a joint effort, and detailed design and execution by the contractor. On the other hand, the client might do all the design. The division of the project as a whole, or the division of the major components into their phases, is often dictated by the characteristics of the project technology. However, before deciding to award contracts or subcontracts always ask yourself, 'What will this other organization contribute, or perform better, that your organization could not do themselves? '

The advantages and disadvantages of breaking up a project into different phase combinations are listed below with reference to Figure 3.2. The names of the different phases are for illustrative purposes – different business or industry contexts will have more phases and use different terminology.

In arrangement A we have decided to let the whole contract to a single supplier or contractor. The advantages of the single contract arrangement are:

- full and clear accountability;
- it minimizes use of client resources;
- project execution is more straightforward;
- the project will use a single set of integrated systems;
- the interfaces are reduced;
- management costs may be reduced.

---

1   The Association of Project Management Body of Knowledge. See *Additional Sources and Contact Details*.

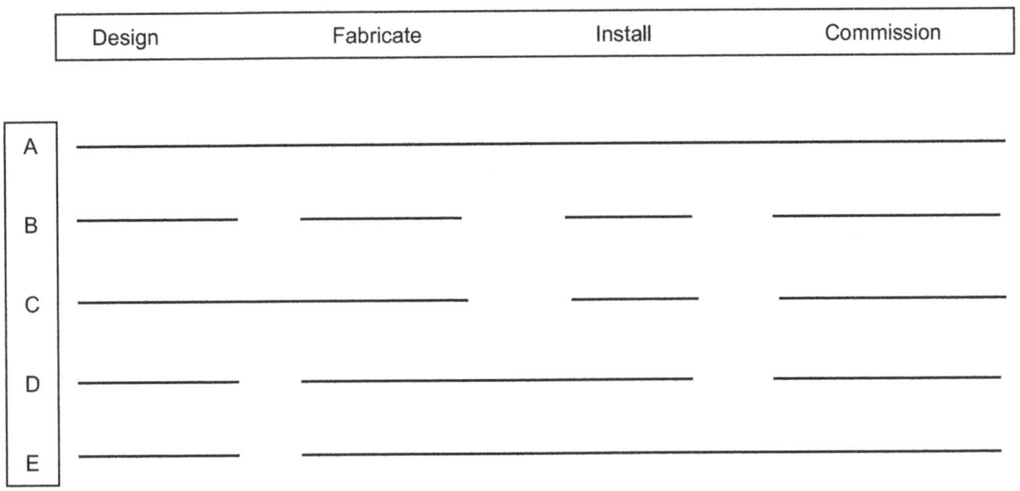

Design	Fabricate	Install	Commission

A ————————————————————————————————————

B ———— ———— ———— ————

C ———————— ———— ————

D ———— ———————— ————

E ———— ————————————

Figure 3.2   Division of the work – options

The disadvantages of single contracts are:

- the supplier or contractor may not have the best expertise throughout all of the phases;
- depending on the size of the project it may restrict the number of tenderers capable of performing the work;
- all our eggs (risks) are in one basket. If the supplier or contractor fails then our project fails.

It is this last disadvantage that might lead us to a more cautious approach and a decision to spread the risks across a number of suppliers or contractors. The advantages of multiple contracts shown in arrangement B are:

- the project can be developed progressively;
- the expertise for each element can be maximized;
- more suppliers or contractors will be able to tender providing a more competitive arrangement;
- smaller and local contractors are able to be involved;
- it might be possible to overlap the phases and achieve a shortage schedule;
- the project risks have been spread amongst a number of suppliers or contractors.

The disadvantages of multiple contracts are:

- with more suppliers or contractors more management resources will be required;
- it may be more expensive with each supplier or contractor applying overheads and profit to each element;
- each supplier or contractor is only accountable for their portion of the contract. The client must take overall accountability;
- new risks have been introduced at the interfaces.

A comparison[2] of a single and a multiple contracting approach, using as similar as possible North Sea projects, indicated that there was no cost advantage to either method. This was valid

---

2   Unpublished by Anton de Witt presented at Cranfield School of Management.

provided the client's management costs were not taken into account; otherwise the single contract method appears to be more cost effective.

The last disadvantage of multiple contracts gives us reason to pause. In arrangement A all our risks were with one person and, because we perceived this to be too big a risk, we spread it amongst a number of other suppliers. However, we have now introduced more risk but in a different form.

A much maligned individual, who was a friend of Leonardo da Vinci, studied people whilst Leonardo studied nature. This individual, Niccolo Machiavelli, wrote an excellent little book called 'The Prince'. If you change armies into project teams and princes into project managers the book provides excellent advice on project and risk management.

> No government [project] should ever imagine that it can always adopt a safe course; rather it should regard all possible courses of action as risky. This is the way things are: whenever one tries to escape one danger one runs into another. Prudence consists in being able to assess the nature of a particular threat and in accepting the lesser evil.[3]

This is a fundamental quotation on Risk Management. What it says is that risk does not go away. You may change the shape and nature of the risk but you then create another risk. This is, in effect, the essence of contract strategy. How can the risks be cut, carved up and moulded to make them acceptable to someone else?

Separating design and fabrication has created a situation where the fabricator will complain to the client about every little error in the design. The client is then faced with a long list of extras that they must try and back-charge to the designer. Since this is not good use of the client's management time it is best to combine design and fabrication as per arrangement C. A typical example of this is steelwork where design and fabrication is placed with a single supplier, and for similar reasons design fabrication and erection is common. Combining design and fabrication is also the norm for all types of mechanical equipment from pumps to fridges.

In arrangement D the process plant client or contractor might do the design of, say, the pipework or electrics but get another supplier or contractor to supply, fabricate and install the materials.

Arrangement E is a typical civil and building arrangement. An architect does the design that is then passed to a contractor, who has had no involvement in the earlier phase, the reason for problems with this type of work. Arrangement E is also an example of any client who has proprietary know-how. They are not going to release the secrets behind their basic technology to a contractor. Accordingly, they develop a basic design package for a contractor to develop and construct.

PERFORMANCE OF THE WORK

There are two aspects to the performance of the work:

i) The division of risk between the contracting parties, that is, the contract category.
ii) What should be done in-house and what should be let out and purchased. This involves the organization of the work.

These two issues are interrelated. When we look at dividing up the work that will be done by ourselves in-house or the work that will be let out, we are, in fact, dividing up the allocation of risk. The two issues are also directly influenced by the level of supervision effort that the client is prepared to make, or capable of making, and has a direct impact on the key decision regarding the choice of

---

3   *The Prince*, Machiavelli, N., translated by Bull, G., Penguin Books, 1961, p 123.

contract category. There is no point in a client saying, 'I will transfer all of the risks to the supplier or contractor,' if the client does not have the skills and expertise to fully and comprehensively describe the scope of work or services. Similarly, it is not sensible for this client to choose a reimbursable form of contract if they do not have the skills and resources to manage and control the contract. This is the dilemma of the naive client lacking experience, expertise and understanding. They cannot have a Fixed Price Contract or a Reimbursable Contract and, consequently, they need to employ someone to act on their behalf. However, does the client have the competence to employ and manage this consultant or management contractor? On one occasion my organization was acting in this role, for and on behalf of the owner, only to discover that the client had also hired another person to check on me! Little fleas on the backs of bigger fleas on the back of… and so on!

Typical management concepts for handling what should be done in-house or what should be let out are:

- A contractor is given full responsibility for all the design, supply and installation activities so as to provide a fully operating facility; a turnkey project.
- A traditional architect or consulting engineer performs the management supervision on behalf of the owner.
- A contractor is employed to carry out the construction management.
- A contractor is employed to supplement the client's capabilities by providing project management services.
- The owner performs the project management.

Supervision effort

How the level of effort and client involvement determines the division of work and influences the type of contract is shown in Figure 3.3.

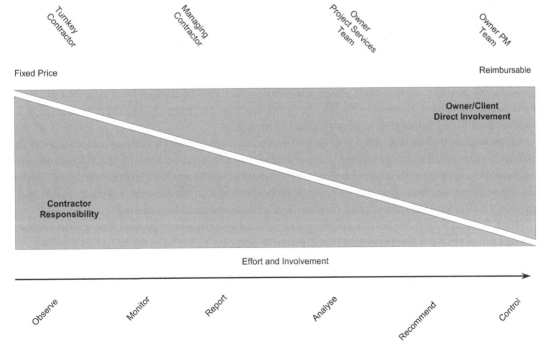

Figure 3.3   Client involvement

If a client does not have the expertise to manage a project and would like to pass all responsibility to a contractor, then they can only observe until the project is complete. In this situation a fixed price contract is possible. However, the client had better not interfere, for once the client gets involved, makes changes or alters the work sequence, the client loses control. The supplier or contractor is then justified in making claims for extra cost for the disturbance to their optimum work method.

Perhaps a client is uncomfortable parting with their tens of thousands or millions of pounds or dollars and would like to have someone keep an eye on the contract and report back on the contractor's performance. As a consequence, a management contractor will be necessary to perform part of the client's function in the project. Since it may not be possible to determine how much work the management contractor will perform on the client's behalf, this portion of work will need to be on a reimbursable cost basis. Where the management function does not need to be involved, the work can be let on a fixed price basis.

The client's confidence and expertise has now increased to the extent that they wish to be more involved. They want to analyze the work methods and comment on and recommend courses of action. In this situation, being a client that runs a different type of business rather than projects, they will need additional resources to perform the project management functions and will need to hire a project services team. More of the project work is now on a reimbursable cost basis and less will be allocated to a contractor to manage on a fixed price basis.

Finally, as a sophisticated client, with resources and expertise, the decision is made to take control of the project and use a team of people from within their own company to manage it. In this situation the client will award a reimbursable contract to a supplier or contractor to perform work as directed by the client.

In traversing this spectrum (in reverse) we can see that as you pass risk you pass control. Consequently, the primary strategic question should be, 'Where do you want to be on the effort and involvement spectrum?'

Contracting map

Having decided on what should be done in-house and what should be let out, objectives must be identified and allocated, for each of the main components or subsections, in terms of cost and schedule confidence limits. At the same time, the appropriate contract category must be decided and all of the contracts mapped in block form, as shown in the example in Figure 3.4.

As far as possible, at this early stage, the map should also indicate which type of contractor is envisaged for which elements of work. Each contract on the map should then be analyzed for interface problems that might occur with other contracts.

Let us assume that the subunit 1 contractor has a fixed price contract for design and a reimbursable contract for the installation or construction element. Being a sharp contractor their proposal indicated that small bore piping (without having defined this term) would be site run. So that instead of 2 inches and below being site run, they site run, say, 4 inches and below. They have, in effect, transferred a significant amount of their fixed price contract work into reimbursable cost work.

The following is an example of contracting strategies for a pharmaceutical plant. The scenario is based on a real project but has been presented for illustrative purposes rather than for historical accuracy.

CONTRACT STRATEGY CASE

A large pharmaceutical company has carried out a number of projects over the years, all of which they consider to have been successful. They have decided to invest in a major science research park and at the same time centralize their administration buildings. The research facilities will comprise at

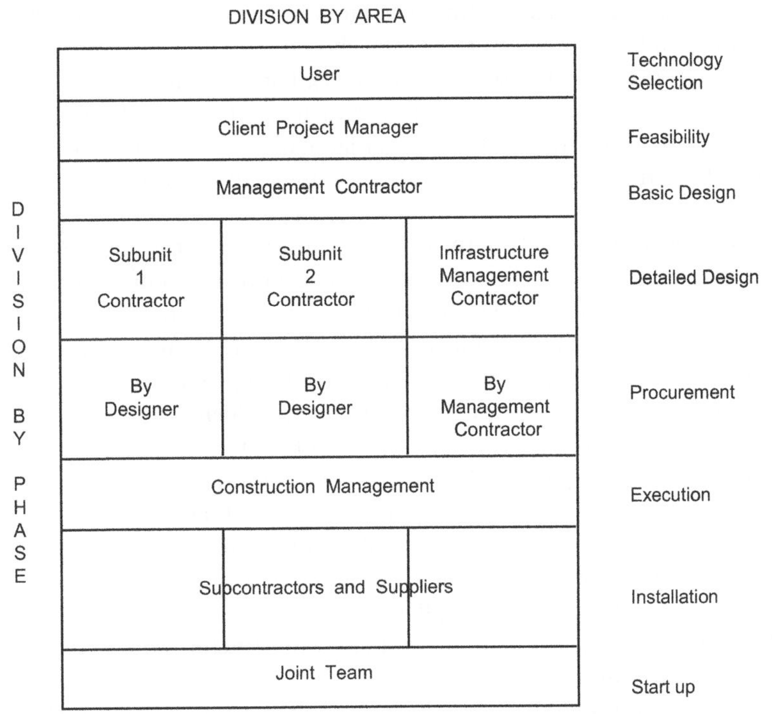

DIVISION BY AREA

User				Technology Selection
Client Project Manager				Feasibility
Management Contractor				Basic Design
Subunit 1 Contractor	Subunit 2 Contractor	Infrastructure Management Contractor		Detailed Design
By Designer	By Designer	By Management Contractor		Procurement
Construction Management				Execution
Subcontractors and Suppliers				Installation
Joint Team				Start up

(left margin: DIVISION BY PHASE)

Figure 3.4   Contracting map

least four high technology laboratory buildings incorporating all the latest technology and a highly flexible pilot plant. The buildings will have one floor of laboratories and one floor of services on an alternating basis. A state-of-the-art biotechnology facility, which is needed urgently, will incorporate clean rooms where all the services are hidden behind walls. A central feature will be the main library, restaurant, reading and meeting rooms positioned on the edge of an enlarged natural lake. The total complex will be spread over a large landscaped area and is likely to cost in excess of £500m.

The company has a senior project manager who has been with them for over 12 years and who is about to finish a laboratory block extension project. They recently transferred out of one of the operations departments to be one of the company's first full time project managers.

How should the company proceed?

Discussion and analysis of case

Do not try to manage a large project, but split it up into the natural subsections at level 2 of the product breakdown structure:

- four laboratories;
- pilot plant;
- biotechnology facility;
- main building, landscaping and site infrastructure.

It should be noted that each one of these is a different technology and needs contractors with different experiences. The mistake would be to let it all to one contractor. Having four different contracts is not an issue here since each one can be separated as an individual project with little or no interdependence.

The laboratories are very high technology building services bordering on an engineering construction culture, and a contractor needs to be chosen accordingly. Consequently, mitigate the risks and award the laboratories as a series of contracts. For example, award one contract and add subsequent laboratories as change orders if the contractor performs well. Alternatively, award one contract and have a second contractor for another laboratory in order to provide competition and an incentive. Tell the contractors that the best performing contractor will be awarded the next two laboratories.

The flexible pilot plant is a typical pharmaceutical plant and is best awarded to a process contractor with experience in that technology. The contract can be let on any contract basis depending on the level of scope definition, and the degree of client involvement required.

The biotechnology facility has a number of problems. State-of-the-art means having the latest technology with the danger of the client wanting to make scope changes. Clean rooms mean high quality workmanship and special construction techniques. Hidden services mean that access will be restricted so that one cannot throw resources at the problem in order to shorten the schedule. And, it is all wanted urgently! Consequently, the best way out of these incompatibilities is to modularize the walls and prefabricate them off site in a factory environment. Even though it might be tempting to split the contract in a similar manner to the laboratories, we need to maximize the benefits of the learning process on the first wall. Further, we would not want an argument over interface errors when matching up units from different suppliers.

Finally, the main building, landscaping and site infrastructure is a traditional architect and building contractor arrangement.

The problem comes, as it did in reality, when one contractor falls behind schedule. The project was being built at the same time as the Sizewell B nuclear power station and the Channel Tunnel so that there was a shortage of labour in the marketplace. Consequently, the contractor with schedule problems tried to attract labour to the site by increasing the wage rate. So where did the labour come from? The other projects on the same site! Therefore, the next contractor with problems had to... and so on, and warfare broke out. Thus, the projects may have been different technologies, necessitating contractors with different cultures, but they were not separate – as indicated above. The contracts needed a common site agreement for the labour employment policy.

This multi-project scenario has many parallels – for example, the labour issues – to the London 2012 Olympics project.

The likelihood is that the company will choose its own project manager. It is only natural. It is their money and their employee and, in the eyes of a naive (in a contracting situation) management, they have been successful so far. This is what happened in real life and it was a tragedy. They did not have the experience to manage a project of this nature. What was needed was a project director with four project managers with the appropriate experience in the different technologies. The cost escalated to over £750m and was late. The consequence was that the only option left was to cut the scope, resulting in a project that did not achieve its original objectives.

This last part has been included to illustrate that the contracting strategy is not just about how the contracts are organized, but it is also about the knowledge and experience of the project management team and how it is structured.

Finally, the contract strategy is not complete until the point of delivery and the payment terms have been considered. Delivery is partly discussed in Chapter 5, but payment has been left until Chapter 10.

4 *Contract Categories*

This chapter takes a less conventional approach to describing contracts, with the emphasis on risk. Unfortunately much of the literature describing contract types is misleading and some erroneous. The existing literature places great emphasis on describing contracts by their broad payment terms and sometimes by how they are organized. Whilst the payment terms are convenient labels to distinguish one contract from another, the broad terms used can be misleading. They do not identify the risk category. When contracts are viewed from a risk perspective they fall into only three categories.

In contracts, risk is expressed by who suffers the financial consequences of a risk event or performance failure. That is, who is responsible, who is liable and who will be paying when matters do not turn out as planned. The conventional contract names, in common use, do not identify the fine points of the payment terms in sufficient detail to correctly identify the allocation of risk (see Chapter 10). Thus, it cannot be overemphasized that *the true nature of a contract is determined by the 'rules' – the words in the contract terms, and not by the convenient labels that are in common usage.*

THE THREE CONTRACT CATEGORIES

At one end of the contract spectrum the contractor provides services to perform work with minimal or no definition. The contractor is reimbursed the cost of the work carried out, but accepts no responsibility (risks) – a reimbursable or cost plus profit basis for the contract. It is clear that if the client pays all of the cost of executing the work then they are the ones who are carrying the risks (technical, commercial and financial).

Tenders for Reimbursable Contracts are evaluated on the means that the tenderers propose to use to execute the project. How effective and efficient are their procedures? Do they have a good execution plan? How well are they organized? Is it a realistic schedule? In effect these are Method Based Contracts.

At the other end of the contract spectrum the supplier or contractor supposedly accepts the liabilities (risks) involved in delivering a fully defined end result or product, for a fixed price. Tenders will be evaluated on how well the product meets the client's requirements and how well it will perform. A Performance Based Contract.

In this situation the supplier's or contractor's price does not change for the agreed scope. If the contractor's prices are fixed (for the given work scope) then the contractor is taking a risk. A Fixed Price Contract can be described as a Contractor Risk Contract. However, this is only partly true and the analysis of this risk is clarified towards the end of this chapter.

These are the two ends of the contract spectrum and, in between, there are a series of contracts where the client takes a risk on the scope of work or services to be supplied but the contractor takes the commercial risks. These contract types are our third category, namely, Shared Risk.

The following is a generic list of most of the contract types in common usage grouped according to risk category and graduated according to the perceived risk to the client. These names identifying payment terms are descriptions for defining the more detailed secondary level of apportioning risk.

However, it must be remembered that every industry or business sector will have its own particular terminology. Even though the specific name of a contract may appear to be a different type it will still fall into one of the three basic risk categories.

Client risk:	cost + % fee
	cost + fixed fee
	time & materials
Shared risk:	dayworks
	unit rates
	bills of quantity
	remeasure
	target cost
Contractor managed risk:	guaranteed maximum
	firm price
	lump sum
	fixed price

It seems that these variants have been developed over time because contractors have learned how to cheat! Or, should I say, get around the intent of a particular contract. However, I contend that the payment terms encourage a contractor to behave in a manner that is contrary to the objectives and best interests of the client. 'Tell me how someone is rewarded or paid and I will tell you how they will behave.' The conclusion from this is that the contracting strategy should be chosen to provide the desired behavioural culture.

The contract extremes produce two quite different cultures. If the client takes the risk with the contractor providing a service, the client can develop a 'let us work together' culture. On the other hand, if the contractor takes a risk and is responsible for delivering an end product, the contractor is more likely to distance themselves with a 'leave me alone to get on with the work' culture. When people are moved (on to the next project) into the opposite culture it produces behaviour that is significantly disruptive and capable of compromising the project objectives. My experience is that, unless specific actions are taken, it can take the life of one project to change peoples' behaviour from one culture to another.

The basic characteristics are summarized in Figure 4.1. The arrows indicate that a particular characteristic is maximized in the direction of the arrow.

The individual contract types within the three risk categories are now described in detail. 'The Introductory Notes of The Institution of Chemical Engineers, Model Form of Conditions of Contract for Process Plants', provides an excellent comparison between the two basic types of contract. Some of their comments are integrated into the following explanations. In one or two areas there may be slight repetition of what has already been covered but it is included for the sake of completeness.

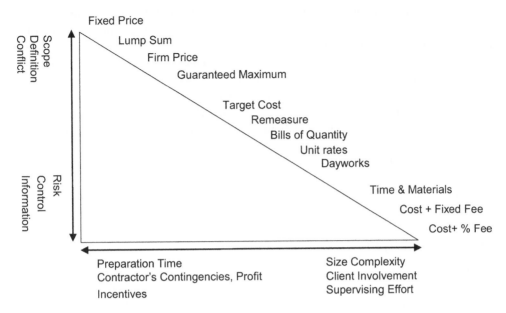

Figure 4.1   Contract characteristics

CLIENT RISK CONTRACTS

Reimbursable contracts

The contractor may be responsible for all aspects of schedule, cost and quality control, but if they are reimbursed at an agreed rate for all manhours used then the client is in effect assuming all of the risk. Thus the contractor's loss is limited. However, their profit is also limited.

Market conditions may be such that contractors are not willing to accept the risks associated with fixed price contracts. For example, for very large or complex projects or during periods of high inflation. In uncertain market conditions the purchaser may consider that contractors' allowances for these uncertainties are likely to be too high and will decide to carry this risk themselves.

A disadvantage to reimbursable contracts is that the purchaser is uncertain of the final project costs. However, once the scope has been fully defined and the specifications agreed, the client could ask the contractor to convert the remaining portions of the project into fixed or firm prices. See Convertible Contract later.

The primary advantage of the reimbursable type of contract is that it permits a rapid start of work before some elements of the project have been decided. The owner can involve the contractor earlier in the project development process in order to evaluate alternative schemes for the project. The rapid start to the project development process can also enable the owner to achieve shorter schedules. It also provides the client with the greatest control over project execution. However, the client must control their desire to be excessively involved, which would result in over-design. On the other hand, the project may be dependent on the technology know-how of the client, or there may be issues that are of a confidential nature, and the client may not wish to issue all of the project details in an enquiry.

This contract category also provides the client with the flexibility to make changes. Unfortunately, for the client, this can be a disadvantage. They may interfere to such an extent that the project cost and schedule objectives are compromised. A major advantage for the client is that they can easily terminate the contract at little cost to themselves.

If a reimbursable contract is to operate successfully the client and contractor must behave and work together effectively. Further, all costs, whether for labour, construction equipment, tools, supplies or other categories, must meet the audit test of need. For an expenditure to be made, the question of need must be satisfied. The authority to determine need depends on the contract terms. If the contract states clearly that the contractor has a certain scope and is deemed to be an independent contractor, the authority to make such decisions belongs to the contractor. For instance, if the contractor is held responsible for meeting the schedule, then the contractor must have the authority to decide how much labour, tools and supplies, construction equipment and supervision is required.

When a client grants this authority and the risks that go with it to a contractor, the latter has more freedom to act; but the client needs a way to require the contractor to adhere to their commitments. This can entail encouraging and motivating the contractor to spend the minimum amount of money to execute the project within the allotted timeframe (see Chapter 12, Incentives).

Cost plus percentage fee

This fee or, in effect, profit arrangement should not be considered as a serious option for the contracting business. However, my research shows that one or two of them still appear from time to time. It must be remembered that this form of contract is very common for the provision of other services, for example, architects, design and cost consultants and lawyers, since it allows the fullest participation of the client and is completely flexible. Nowadays a professional firm will provide an upfront estimate of these fees so that the final costs or fees are not a complete surprise.

In theory, any conflict between client and contractor should be minimized. However, the behaviour of the contractor can be contrary to the client's objectives. The contractor has little, if any, incentive to meet cost and schedule budgets and is actually rewarded for overruns caused by mismanagement or over-design. It does not take long for the contractor to realize that they cannot lose. 'Delighted to work with you,' they say, 'tell us what to do and we will provide our top designers or other specialists who will use their very best skills.' Here is an opportunity for the contractor to try out new design technology and develop new systems, and the client pays all of the costs with a guaranteed percentage profit. The contractor will want to look at different options, different schemes and different methods, checking their work carefully, since the more work they do the more money they make. In order to keep control of the work the client expends maximum effort and needs to be highly competent using large numbers of people in checking and auditing the work of the contractor so as to avoid and manage the risks involved. Not a satisfactory contracting process! The classic example of this contract format was the Nimrod AWACS project, where in the Ministry of Defence's documentation it stated that, 'The contractor is insulated from loss.'

To provide a more balanced view it should be stated that this behaviour depends on the nature of the contractor. Most reputable contractors would, or should, look after their clients' money as effectively on a reimbursable contract as their own money on a fixed price contract.

The main disadvantage of the lack of motivation to meet cost and schedule targets can be overcome by means of a bonus incentive scheme. A separate incentive scheme enables the contractor to enhance their profit by beating mutually agreed goals on schedule, fabrication progress and safety standards.

The mismatch in objectives and behaviour can also be overcome by changing the contract payment terms. The first step might be to provide a sliding scale for the fee – the more the costs increase the less the fee percentage. Ultimately, the most sensible action is to change, from a percentage profit or fee, to a fixed fee.

Cost plus fixed fee

This contract has most of the disadvantages of the percentage fee contract, but in this contract type the fee is fixed, and can be amended to cover overheads as well as profit. The fixed fee, once established, does not vary with the actual cost. However, it can generate considerable controversy when 'perceived' changes are made to the original 'implied' scope on which the fixed fee was based.

Changing the payment terms changes the behaviour of the contractor. Although the contractor is not rewarded for overruns there is a small incentive for under runs. For example, the contractor quotes, say, 1000 hours at £2 per hour margin. Suppose they manage the work inefficiently and it takes 2000 hours. The margin is now reduced to £1 per hour. 'So,' says the contractor, 'let's get on and finish this work since we can sell our manhours in the marketplace for £2 per hour margin.'

In the Cost Plus Percentage Fee and Cost Plus Fixed Fee the non-technical administrative cost are likely to be high, since all costs are being paid for. The next step, therefore, is to fix the fee to cover all non-technical activities.

Direct cost reimbursable plus fixed fee for indirect costs

In reality the only resources that the purchaser needs and wishes to use and direct are the technical specialists. This has the benefit of not only making a contractor become more efficient with the support functions but it also reduces the client's involvement and auditing efforts. As a result, the controversy over changes can be even more fiercely fought.

Except for the pure cost plus percentage fee contract all reimbursable contracts have elements that are fixed. It can, therefore, be misleading to talk about a 'reimbursable contract'.

All cost reimbursable contracts are sometimes termed Time and Materials Contracts. However, this is a rather broad and imprecise term. Which of the elements discussed above are included in the time rate? Is the rate cost based or price based? It is perhaps appropriate here to remind ourselves of the meaning of these two words:

- Costs are a matter of fact. There are records to support what has been spent on the various work activities, and receipts to justify how much has been paid for materials.
- Price is what you can sell your goods or services for in the marketplace. We should also remember that a supplier's or contractor's price is a client's cost.

SHARED RISK CONTRACTS

For the client to reduce the risks associated with reimbursability they need to do some design work. They need to identify the basic elements of work involved in a project. The contractor can then be asked to provide some rates – that is, prices for all of the activities required per unit of measurement for, say, cable laying, casting concrete, plastering walls and installing fittings, as well as prices for all of the materials to be used. When the work is finished the client can remeasure what has been done and identify the quantities of materials used, and consequently, pay the contractor accordingly.

Unit rate contracts

These are very common contract types and variants of them have been, and are, the civil and building traditional way of working. In this type of contract the actual quantities to be installed determine the cost or price of each tendered item of work. The contract is tendered on the basis of estimated quantities but the quantities are often difficult, if not impossible, for the client to estimate accurately before contract award. The rates quoted can include allowances for overheads and a profit element so that they can be, in effect, prices.

Obviously, the client assumes the risk of quantity variations and the contractor the commercial risks for the pricing of work activities and the cost of materials. As a result, these are a Shared Risk Contract category.

Shared risk contracts place an emphasis on one particular aspect or another. A schedule of rates for home office work, say, and the client pays the cost of materials or a fully developed listing of material quantities and a contractor prices for their supply and installation (a Bill of Quantities.) In all of these mechanisms they are, in essence, dividing up the risks associated with the project.

To arrive at the final contract figures the work has to be remeasured. However, the remeasurement of the work done can be abused by, say, the contractor digging too big a hole, pouring too much concrete or installing unnecessary valves. So remeasurement should only be carried out on drawings showing what should have been done, not what is actually done. This form of abuse is less likely to occur with mechanical or electrical work activities.

The danger with rates type contracts is that both client and contractor might play contractual games. Perhaps a client perceives one section of work to be overpriced and, consequently, issues a change deleting it from the scope of work. The consequence being that the contractor may not recover the overhead and profit that they had anticipated. Alternatively, the contractor may load a particular rate for an identified small quantity of work, say rock removal, which being one rate amongst many for a particular section of work may not be significant. However, the contractor knowing the site conditions in more detail is aware that there is actually a large quantity to be removed and, consequently, the client ends up paying significantly over the odds. Obviously, the contract needs to provide for protection to both parties if actual quantities vary significantly from those tendered in the schedule.

Further, the contractor may 'front end' load the rates inflating their overhead and profit portion of the rate for activities that will be done early. This subject is discussed in detail in Chapter 12.

One Dutch client of mine used what they termed a Boxer (BOQSOR) Contract – a Bill of Quantities/ Schedule of Rates. With this level of definition one should consider using a fixed price contract. On one occasion when I was visiting the Civil Service College to lecture on project management I noticed, in the entrance hall, a full mock-up of a study bedroom for a proposed hall of residence. There was also a set of architect's drawings showing plans and elevations. When I cross-examined the project sponsor, who was attending the course I was lecturing on, I was surprised to learn that there was also a full bill of quantities that they proposed to use for tendering purposes. With total definition, why not pass the risk for the quantities on to the contractor and ask for a fixed price?

Target cost contract

Strictly speaking this is not a separate contract type since it is essentially a Rate Based Contract with a built-in incentive mechanism. The rates are cost based and there is a separate fixed overhead and profit element. The objective of this type of contract is that the less the contractor spends the more they earn. Usually the incentive basis is that any under or overruns are shared equally between the client and the contractor. The intent is to provide an incentive to retain the contractor's interest in the contract since there is a potential to earn more profit. This idea worked for a while until contractors realized that, every time the client made a change, there was a good argument in favour of convincing the client that the target should be raised. After all, if the additional work had been known about at the start of the project the target would have been set higher. We now have a situation where the client and contractor's objectives have diverged. The client is trying to reduce their costs whereas the contractor is trying to push up the target so that they can always be below the target and share a saving. The consequence of this is that, initially, clients abandoned this contract type and requested a guaranteed maximum price (see later). How to make this contract mechanism work is discussed in Chapter 12, Incentives.

CONTRACTOR MANAGED RISK CONTRACTS

The more of the work performance risk that a purchaser wishes to pass to a contractor, the more preparatory work needs to be carried out. So that, eventually, there is a full set of design drawings, a comprehensive schedule of rates and a fully developed list or bill of materials. As indicated above, with comprehensive definition of the scope, it is sensible to choose a different contract form and save all the measurement effort and auditing of the costs involved. This saving is, of course, offset by the cost of preparing detailed tender documents. However, does a purchaser take any of this into account when choosing a contract category?

In theory, the purchaser knows their expenditure commitment. But, as we all know, no client can stop themselves from making changes (after all their business needs will change over time) – so do they know what their costs will be?

Contractor Managed Risk Contracts may be perceived to be more expensive owing to higher allowances for risk and escalation. There will also be higher profit margins to compensate the contractor for the risks that they are taking. This raises the question: Is it better to pay the actual cost of what actually happens (a reimbursable contract) or is it better to ask someone to estimate what might happen (a Fixed or Firm Contract) and allow them to keep the extra money if they manage the risks effectively?

As previously indicated, it should be evident that it is essential for the purchaser to provide complete definition of the scope together with detailed specifications, as a result of which changes are more easily identified. Since the project will most likely have been tendered competitively, the contractor will be looking for ways to recoup margins particularly if the client has awarded the contract on a lowest price basis. The client may ask innocently, 'What about this x part of the scope?' To which the contractor replies, 'Ah, that will be extra.' 'No,' says the client, 'it's included in the specification,' – to be greeted with, 'No, if you look at the words in the specification or, in our tender, you will see that it is excluded.'

Contractors tend to be risk averse and avoid the risks associated with a fixed price. They should, however, like this type of contract; for if they manage it well the money saved is additional profit. However, the consequence for the client is that the contractor will under design in order to reduce their own costs. Thus, the client must be sure that the product produced performs in accordance with the specification.

This contract category is most commonly used for the purchase of materials or commodities from suppliers. It is much less likely to exist in its pure form (fixed price) for a project consisting entirely of services – there will always be uncertainties and risks where a contractor will prefer to be reimbursed their costs or where the client is unable to be precise about the scope, specification or quantities. It is, however, now possible to get law firms to competitively tender fixed prices for conveyancing, since the scope of this service has now been agreed nationally.

The most appropriate use of a fixed price contract for design and fabrication (for example, equipment and steelwork), or for design and build (for example, civil work), or for engineering and construction (for example, power and process industries), is when a supplier or contractor has their own proprietary design or process. Another effective use of this contract category is when the purchaser has completed all of the basic design work. In these situations the purchaser should allow the suppliers or contractors full rein to exercise their know-how competitively.

As a result, the general nature of the contract precludes any client involvement without it costing them money. The client gets involved at their peril. The result is confrontation and conflict. Not a satisfactory contracting process!

There is no sharing of the management of the risks and, in theory, the supplier or contractor is responsible for the risks. However, the project description provided later demonstrates the fallacy of this point of view.

A guaranteed maximum cost or price

This contract is essentially a Cost Based Rate Contract with a fixed unalterable cost target with a separate fixed overhead and profit element. The under runs are shared in the same way as for a Target Cost Contract but the overruns are 100 per cent to the contractors account. In effect, a fixed price contract with, at best, 50 per cent of the cost savings as profit. On a fixed price contract the profit margin in the price is enhanced by 100 per cent of any savings! For the contractor this is a high risk contract. The contract is put together in a cooperative rates environment and yet it has more financial risk for the contractor than a fixed price contract.

Fixed price

My definition of this contract type relies on the meaning of words. Fixed (for a given scope) means fixed, it does not change. However, some business and industry contexts interpret this slightly differently. Yes, the rates are fixed to a scale of charges, and if the marketplace for this basket of prices changes then the rates are adjusted accordingly. They have used the term fixed price in the same manner that I now use the term firm price.

Firm price

A Firm Price Contract has most of the issues of the fixed price contract but in this contract type the price is fixed for everything except some item for which a contractor is unwilling to take the risk. This might be overall inflation or it might be the risk of National Insurance contributions rising. A typical example would be inflation on the price of a particular commodity. An excuse given for part of the increase in costs for the London 2012 Olympics was a supposed doubling in the price of steel. In these circumstances there should be a Cost Price Adjustment (CPA) clause in the contract.

Thus, except for the true wholly fixed price contract, all other types of contract involve a greater or lesser degree of reimbursability.

Lump sum

In the last 10 years I have found many instances where the term Lump Sum Contract is used to describe a fixed or firm price contract with a 100 per cent payment in a lump sum at the end of the project. It is how many materials or commodities are purchased. Since it applies to either a fixed or firm price contract, it may be seen as just a payment mechanism and not a different type of contract. However, it certainly increases the supplier's or contractor's risk, and encourages them to perform effectively until the client has accepted delivery of a completed product. If only more people who contract in the home improvement market would use this mechanism.

If we rely on the meaning of words I think the description of firm price and lump sum, I have used above, should be the preferred ones. I cannot emphasize too strongly that one should never rely on a specific definition of these words; different business and industry contexts use these terms in different ways. Always ask the other party what they understand by the terms. Since contracting will almost invariably cross business and industry cultural boundaries, the other party is very likely to understand the terms differently. Sometimes the term Fixed and Firm price is used. I have never been wholly sure why people use this term unless it is to imply that some costs are fixed and some costs are subject to escalation, that is, there is some reimbursability. Apart from the pure fixed price contract and the pure reimbursable cost contract at the two ends of the contract spectrum, this is true for all contracts.

All other Contractor Managed Risk Contracts contain some elements that are reimbursable, and all other Client Risk Contracts contain elements that are fixed.

Another expression often used is a Convertible Contract. This is not a separate type of contract since any contract can, by mutual agreement, be negotiated from one type into another type of contract. The expression is usually used for an agreement to convert, at some appropriate stage, the client risk reimbursable contract to a contractor managed risk contract. The conversion is likely to occur at some midpoint in the development of the project concerned, when the scope has been fully defined. Naturally the contractor will perceive that there are still some risks remaining in the project and will include contingency allowances accordingly. The client, however, knowing the contractor's costs in detail, from the reimbursable rates that the contractor has supplied, can evaluate these contingencies. As a result, the client can be surprised by the price submitted and conflict can arise over what is seen as extra profit. This conversion is, in effect, a single tender offer. Consequently, the client needs to decide whether to allow the costs to be reimbursed or treat them as non-reimbursable tendering costs.

Finally

There are other contract expressions in common usage that are not contract categories or types, and these expressions are explained in the next chapter.

PROJECT CASE STUDY

The following project story provides the reason why so-called contractor risk contracts should be renamed contractor *managed* risk contracts – particularly viewing the contract from the owner's perspective.

For many years Exxon stated that they believed in fixed price (their term lump sum) contracting and in 1987 an article[1] by a section head in the contracting and project services division of Exxon Research and Engineering's project management unit stated:

Owners, such as Exxon, prefer lump sum contracting because it establishes better definition of project cost and reduces financial risk.

However, prior to 1983 in the UK, they let nothing but reimbursable contracts.

With the introduction of lead-free fuel Exxon needed to build a gasoline unit on their refinery at Fawley in the UK. The enquiry was issued in 1983 in Exxon's usual professional fashion for a reimbursable contract. Negotiations were down to the last two tenderers and, just days before signing a contract, Exxon decided to re-evaluate their investment plans worldwide. With Exxon this means everything stops immediately and the project was frozen. The Exxon enquiry team knew that this was a business critical investment and that it would eventually get the go ahead. They, therefore, revamped the project to meet the revised business strategy – in practice this meant that

---

1   'Lump sum contracting an owner's perspective' by Dwight R. Johnson, Jr., *Cost Engineering*, Vol 29/No.2, February, 1987.

the project was little changed. However, the enquiry team used the time constructively and defined the scope and specifications to be used. The enquiry was then issued to the same companies on the tender list but requesting fixed prices and the contract was awarded to the Davy Corporation (at that time the largest British contractor).

The contract was a disaster for, I believe, a fundamental risk management reason. One can manage one of the following, but not two at once. Two at once spells trouble:

- a first of a kind;
- the unfamiliar;
- something new or novel;
- innovation.

The gasoline project had the following characteristics:

- This was the first time Exxon, in the UK, had let a complete major project on a fixed price basis.
- Davy had never worked at Fawley before. The significance of this can be appreciated when one realizes that all engineering construction work in the UK was, at the time, carried out under a National Labour Agreement, except Fawley which had its own agreement.
- Exxon let the contract to Davy even though they knew (Exxon will have had a comprehensive estimating database) that Davy had significantly underpriced the pipework elements – the most significant part of the project.
- The pipework contract was let to Costains.

Any one of these would have been a problem, but all four meant that the project was late! Some suggested reasons for this are:

- After 30 years of running reimbursable contracts the client will not have changed their behaviour and will have interfered seeking innovative ways of executing the project.
- Similarly, the labour force at Fawley will not have taken kindly to a contractor that did not understand their particular ways of working.
- Once the contractor realized that they had underpriced the project, their objective will have been to minimize their losses and not to meet the client's objectives.
- Costain's process division was originally Pertrocarbon, a competent process contractor. So letting the piping to them might have been all right. However, the company had become dominated by civil engineers who did not understand the culture of the process business.

Finally, in frustration Exxon even said to Davy, 'Forget about the fixed price, we will pay you on a reimbursable basis in order to get the project finished.' Even this was unsatisfactory and Davy's contract was terminated. Exxon brought in their tried and tested favourite contractor, Foster Wheeler. Overall the project was 18 months late. So, who carried the biggest risk? Davy? Or Exxon who lost 18 months of the revenue stream!?

SUMMARY

In summary, there are three categories of contract, whether they are viewed from a risk perspective, a money perspective or a behavioural perspective, and this is illustrated in Figure 4.2.

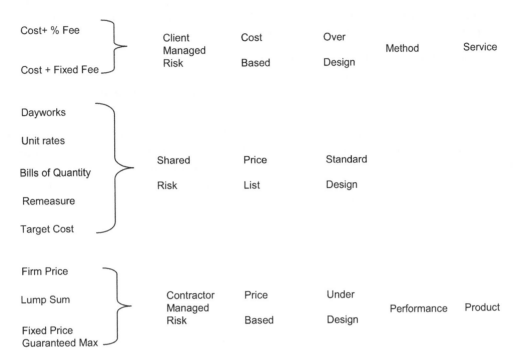

Figure 4.2   Contract categories

Issues to highlight in these categories are:

Risk:
- When you pass risk you pass control.
- In a fixed price contract the supplier or contractor is managing the client's risk and the client should not interfere.
- Due to the contingencies for the risks, the fixed price contract can cost more.

Money:
- Being price based, the fixed price contract will cost what the contractor thinks the market conditions will allow.
- There are significantly more client management costs with reimbursable contracts.

Behaviour:
- A reimbursable cost contract providing a service produces a different behavioural culture to a fixed price contract delivering an end product.
- In a fixed price contract the client must be vigilant to ensure that the product produced meets all of the requirements of the specification.
- In a reimbursable cost contract the client must control their desire to be involved with preferential design work.

Finally, buying equipment for a fixed price, using a performance specification, will clearly create a performance contract. Similarly, contracting for services to design, install, construct or build, on a fixed price basis, will also create a performance based contract.

However, it is easy for a client, with the best will in the world and without any devious intent, to transform a method based contract into a performance contract with any of the following statements:

- The contractor will demonstrate or show...
- The contractor will prove...
- The contractor will perform the work in a diligent, workmanlike and skilful manner and in accordance with best practice.

None of these statements are of concern for the client; they are what the client wants. The contractor, on the other hand, should be concerned. They may believe that they have a reimbursable method based contract, but the above phrases will turn the particular aspects into the contractor's risk. If they cannot meet the required proof, then they may not get paid for that part of the work, and even worse, they may be in breach of contract.

5 *More About Contracting*

FIXED PRICE VERSUS REIMBURSABLE

Before finalizing the appropriate contract category we need to consider the schedule and financial implications. The Exxon article[1] initially referenced in the previous chapter also states:

> *Under competitive market conditions lump sum contracting offers substantial investment savings to the owner. Many contractors, and surprisingly some owners, have disputed this contention. Nonetheless, Exxon has gathered a great deal of data over many years and covering hundreds of projects which clearly demonstrate this point.*

There is an indicative graph, in the article, showing, 'Exxon's own cost estimates defining the zero (or base line),' and tracks 'reimbursable cost projects...and concurrent lump sum projects.' The graph shows that savings of up to 50 per cent seem to have been achieved 'under competitive market conditions' on fixed price (their term – lump sum) projects. Whereas, 'Except for occasional excursions, the reimbursable jobs are brought in close to the estimate levels.' The theme of the article is that, 'Competitive lump sum contracting has been and continues to be the owner's preferred contracting route.'

The conclusion from this is that Fixed Price Contracts should be our first choice. The owner or client saves a lot of money and, as previously mentioned, the contractor can make more money. 'Lump Sum offers the contractors an opportunity to earn money if they estimate sharply and perform well.' To this one should add, provided that the client does not interfere and make changes. 'Whereas...reimbursable cost contracts are at best a break-even venture.' However, it is the peaks – the 'occasional excursions' with the reimbursable contracts - that are of interest (discussed later below).

Having compared cost benefits we now need to compare durations. Figure 5.1 indicates the main differences between the two categories of contract. Client risk contracts can issue Invitations to Tender (also known as ITTs) and be awarded relatively quickly with considerably less documentation. Further, client and contractor can start work almost immediately. However, the project execution phase is longer for the Reimbursable Client Risk Contract, since project definition has to be developed after contract award, giving a project completion at 'A'.

On the other hand, if the client wants the contractor to manage the risks, more documentation is required. Complete scope definition, together with detailed site conditions, is necessary before the

---

1   'Lump sum contracting an owner's perspective' by Dwight R. Johnson, Jr., *Cost Engineering*, Vol 29/No.2, February, 1987.

invitation to tender can be issued. Further, the contractor must be given more time to estimate the project costs and the risks, and arrive at a fixed price. However, the project execution phase is shorter since project definition has already been done, and the contractor will have produced some design work during the tender period. The result is, though, a longer overall project duration at 'B'.

If the client manages the reimbursable cost process well the benefits of the revenue stream are achieved significantly earlier – perhaps up to four time periods. Comparing the net present value of the revenues for the two on-stream dates, A and B, invariably indicates that most benefit is obtained for the client by a reimbursable form of contract. However, we need to examine further the occasional excursion peaks, mentioned earlier.

With a Client Managed Risk Reimbursable Contract, the client and contractor are working closely together. The question is, therefore, how do they behave under these circumstances? Experience shows that two specialists, whether they are designers, planners or project engineers, will disagree! Both parties will be concerned with recognizing, controlling and avoiding risks and, being naturally risk averse, one will want belts and the other will want braces. Being specialists working together they can easily get carried away with preferential engineering, and thus we can get bells and whistles! The result of this is a cost peak over the estimate, and project completion is likely to extend to 'A1'.

On the other hand, with a Contractor Managed Risk Contract the contractor is left alone to get on and do the work. Further, with inspired project management leadership, not only does the contractor make money but completion can be brought forward to 'B1'. More evidence that fixed price contracting should be the choice? No! There is a fallacy in the arguments that have just been put forward. If there can be good leadership and management in one situation, why not in the other? Consequently, the client managed reimbursable contract will always have an overall shorter duration with the resulting earlier generation of the revenue stream and overall financial benefit.

Whilst the article referred to above mentions the cost savings to the client in a fixed price market, it does not mention that some of the savings are offset by the client's additional costs in preparing the enquiry documentation and tender evaluation. However, there are additional client costs involved in managing a reimbursable contract, so perhaps this element balances out.

I believe the earlier completion and revenue stream financial benefit far outweighs the lower prices obtained for fixed price contracting. Further, we must not forget that fixed price contracting precludes any client involvement, and the benefits are only obtained if the project is completed on time.

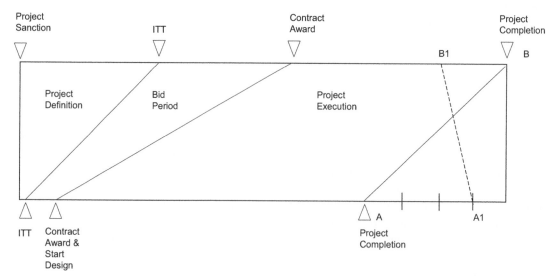

Figure 5.1   Comparison of contract durations

The classic example of causing a Reimbursable Cost Contract to move from A to A1 is the creative architect. To be fair, the trait of focusing on the beauty of the design rather than designing to a budget is applicable to designers in all disciplines, to some degree.

The dilemma of the naive client who cannot have a fixed price contract or a reimbursable contract because they lack experience and expertise was the root cause of the fiasco of the Scottish Parliament building. It took The Holyrood Inquiry by the Rt Hon Lord Fraser of Carmyllie QC 8 months, at considerable cost, to come to questionable conclusions. The outcome of the Scottish Parliament project was clear from the beginning and should have been learnt from the inquiry into the Nimrod AWACS project.

I do not question Mr Campbell's (Counsel to the Inquiry) summing up that the project 'has been characterized by a catalogue of failures, mistakes and incompetence most of which could have been avoided had those in charge exhibited stronger leadership.'[2] I would have said, 'Had everyone involved understood contracts and the contracting process better.' The Inquiry Report by Lord Fraser (15 September 2004) contains the telling phrases that dictated the outcome of the Scottish Parliament project.

It was obvious from the start once the client decided that they wanted a high quality building and to be 'the most important patron of the architecture of government for 300 years... that his heart was in developing a contemporary icon...he wanted to make a landmark building...' using an architect who did not 'think about economy with any seriousness.' Further, as the report's principal conclusions state, 'Whenever there was a conflict between quality and cost, quality was preferred.' The need for urgency was a similar driving force to compound the subsequent problems, 'Whenever there was a conflict between early completion and cost, completion was preferred...' *What happened was bound to follow.*

Lord Fraser places considerable emphasis on construction management being the single most flawed decision of the project (in this case construction management was a client managed risk, reimbursable cost, contract described later in this chapter under 'Scope of Supply'). The implication being that it should not have been used. If you do not spend time upfront defining the scope, then you have little alternative but to define what you want as you go along. The client, their advisors and, I am bold enough to suggest, Lord Fraser never really grasped the fact that they faced the dilemma of fixed price versus reimbursable discussed earlier and illustrated in Figure 5.1.

'The contract was chosen[3] on the basis that it allowed design changes during construction and would allow work to start immediately, thus satisfying Mr Dewar's desire to give Scotland a parliament quickly.' A strategy decision that a client is perfectly at liberty to make but they must also be able to manage the consequences. A fixed price contract would have required the project to be delayed until all the design details had been finalized and agreed by ministers and MSPs. This would, of course, have delayed the start and made the estimate of the final cost more realistic. All of which might mean missing a crucial political window in time and the project might never go ahead.

The report does not consider the issue that if some form of contractor managed risk contract had been used, then any self-respecting contractor would have requested cost and time extensions for specification changes. The net result, though, would still have been the same. It is possible, however, that a fixed price contract using a main contractor would have highlighted the problems earlier. A main contractor, with overall responsibility, would have been disciplined about claiming extras for client changes. This could have made the client more aware of what was happening and a better outcome might have been achieved.

The Scottish Parliament project was a Nimrod project built in granite. The similarities between the projects are uncanny:

- Both replaced existing facilities (old aircraft – old building).

---

2   *The Scotsman*, Wednesday, 26 May, 2004
3   *The Daily Telegraph*, Thursday, September 16, 2004.

- Both manipulated the business decision. 'AWACS more expensive than Nimrod.' The Parliament will only cost £40m.
- Similar projects done before (US AWACS – Berlin Parliament/British Library).
- Both came into being after extensive lobbying.
- Both had the same client (Government).
- Both had similar stakeholders – the RAF and the public and MSPs and the public.
- Both suffered because of *Ego* risk. Buy British – make an architectural statement.
- Both proceeded without full scope definition.
- Both had optimistic schedules.
- Both chose favoured contractors. '[GEC] told us they could do it.' Bovis chosen because of their performance on the Museum of Scotland.
- Both had new elements. New technology and new design.
- Both used reimbursable cost plus type contracts.
- Two separate client committees managed Nimrod. Two separate architect offices manipulated the Scottish Parliament.
- There was a lack of reporting the cost and schedule status on both projects.
- The computers would not fit into the Nimrod airframe – resulting in significant changes. The human computers (MSPs) would not fit into the building framework – resulting in significant changes.
- Both significantly overran budget and schedule.

Quotes from the BBC video 'Bad Deal for Britain' say it all. Change the names and the quotes could apply equally to the Scottish Parliament project.

The Government – MOD-PE[4] chose to buy Nimrod on a cost plus contract. It is now argued that Marconi [GEC] could never have been offered a fixed price contract because of the unknown factors implicit in the development of the radar, yet Marconi must have had some idea of what the project would cost them, and we know the Ministry certainly had their own estimate. However, the contract was offered at cost plus.

…If you have a cost plus basis for a contract, whether it is a development contract or anything else, then you pay for the work which is done, and the work which is done is controlled by that company…

… Some of the extra cost is certainly due to changes in the specification. Nearly every project has changes in the specification…

…The principal development contractor told us that they could achieve something at a certain price, within a certain timescale, and haven't been able to do it.[5]

It is one of the standard conditions of any cost plus contract, and you will find it in the Ministry documentation on the topic, which says that 'the contractor is insulated from loss.[6]

People were upset but you can jump up and down – it doesn't change the price of tea – if you still have work to do to get the thing to work properly.[7]

4   Tom Mangold BBC reporter and presenter of the video, 'Bad Deal for Britain.' Education and Training Video by BBC Enterprises Ltd. 1985.
5   Adam Butler, Minister of State for Defence Procurement, November, 1984.
6   Jack Pateman, Managing Director GEC Avionics – Formerly Marconi Avionics.
7   Sir Douglas Lowe, Controller Aircraft 1975-82. Chief of Defence Procurement, 1982-83.

Lord Fraser in his speech delivering his report ends with his most important words (*not in the report!*), 'A construction management contract costs what it costs.' The equivalent of 'the contractor is insulated from loss'.

Anyone who has had an extension built on to their house knows that an architect likes being creative, and if you make changes and increase the size of a project it will cost you more, regardless of the form of contract.

The difficulties of a naive client (in the above cases the Government) managing reimbursable contracts were learnt in 1986–87[8] The cases emphasize that organizations, businesses or industries seem unable to learn from the experiences of projects in other contexts. *The issues are the same in all projects. The challenge, for the client project manager, is to recognize them and apply them to their own project context.*

Logic, therefore, dictates that unless the project is one where the technology or process is extremely well known and defined, and is not likely to undergo change, the basic relationship between the client and the contractor should be some form of client managed risk, or reimbursable, contract. Fixed prices should be incorporated for items and activities over which the contractor has specific expertise and control (proprietary technology, licence fees, management and clerical functions) since these are determined by how the supplier or contractor runs their business. However, a client's preference will be to maximize the fixed price elements.

As far as risks are concerned, the contractor can sensibly only offer to remedy defective design work or accept penalties up to the limit of their fee. Other risks should be covered by insurance. The imposition of unreasonable risk will only dampen innovation.

The most successful contractual arrangement will be one where responsibilities and risks are clearly defined and allocated between the client and the contractor. Additionally, it will be an arrangement where excessive layers of management are avoided and interference and changes are minimized.

The conclusion is that *clients must learn how to manage the risks and maximize the benefits of reimbursable client risk contracts.*

Incentives

Sometimes people use the term an Incentive Contract. However, it must be emphasized that this is not a type of contract. One can have an incentive bonus scheme on a fixed price contract, or on a reimbursable contract. Further, an incentive scheme should stand alone as a separate contract from the main contract. One can take the view that the nature of a fixed price contract provides its own incentive mechanism. However, the incentive for the contractor or supplier is mainly on the cost elements to the detriment of the client's interest in quality, and there is little schedule incentive.

The subject of incentive mechanisms is discussed in detail in Chapter 12.

OTHER CONTRACT EXPRESSIONS

There are a number of contract expressions in common usage that are often described as separate contract types. For example, a design and build contract, an FOB (Free On Board) contract, an EPC (Engineering Procurement and Construction) contract. It could be a fixed price EPC contract or a reimbursable EPC contract. Thus, these expressions are not separate contract types; they are another aspect to the performance of the work and, as such, are part of contract strategy. They describe the responsibility boundaries to services to be provided by the contract, as tabulated below in Table 5.1.

---

8   House of Commons 6th Report from the Committee of Public Accounts, Session 1986/87 – *Control and Management of the Development of Major Equipment.*

Table 5.1 **Responsibility boundaries**

Point of supply:
Ex. works
FOB (Free on Board)
CIF (Cost, Insurance and Freight)
Scope of supply:
contract/construction management
management contracting
design and build
EPC (Engineering, Procurement and Construction)
EPIC (Engineering, Procurement, Installation and Commissioning)
turnkey
Duration of supply:
term contract
concession

It is useful to summarize how the boundaries expand the responsibilities and liabilities of the supplier or contractor and reduce the buyer's risk, as we progress down Table 5.1.

Initially, the client is responsible for all activities, including the purchase of all materials and collecting them from outside the factory gate. The next step is to ask the supplier or contractor to deliver not only their own materials and or equipment, but to purchase all other materials required and to manage the subcontracts for the execution phases of the project (Management Contracting). In order to reduce the client's involvement and risks further, we can ask the contractor to be responsible for integrating the design and construction phases (Design and Build). This leads us to ask the contractor to provide a facility that is almost ready for operation (EPC) and, eventually, we want a facility that has had the bugs ironed out of it and is operational (Turnkey). Finally, we wish to divest ourselves of as much of the details of the contracting processes as possible, and request the contractor to operate the facility and provide us with the required service for a stated period of time (Term Contract).

Point of supply

FOB, CIF and so on, are 'Incoterms'[9] (see below). They are the, 'ICC (International Chamber of Commerce) official rules for the interpretation of trade terms' applying to the contract of sale. The latest revision came into force on 1st January 2000. In the introduction to this excellent booklet (in both English and French) it states, 'The purpose of Incoterms is to provide a set of international rules for the interpretation of the most commonly used trade terms in foreign trade. Thus, the uncertainties of different interpretations of such terms in different countries can be avoided or at least reduced to a considerable degree.' Each term is defined and the seller's obligations are listed under a series of consistent paragraphs A1 to A10 (for example, A4 delivery, A5 transfer of risks, A6 division of costs, A8 proof of delivery and so on) these are matched on a facing page listing the buyer's obligations in similar paragraphs B1 to B10. The 13 terms are grouped into four categories:

E   Goods available at supplier's works.
F   Supplier delivers goods to carrier appointed by buyer.
C   Supplier appoints carrier and delivers goods to carrier.
D   Supplier delivers goods to country of destination.

---

9   See *Additional Sources and Contact Details.*

Table 5.2   **Incoterms**

EXW	Ex. Works	
FCA	Free Carrier	
FAS	Free Alongside Ship	
FOB	Free On Board	
CFR	Cost and Freight	
CIF	Cost Insurance and Freight	
CPT	Carriage Paid To	
CIP	Carriage and Insurance Paid to	
DAF	Delivered At Frontier	
DES	Delivered Ex Ship	
DEQ	Delivered Ex Quay	
DDU	Delivered Duty Unpaid	
DDP	Delivered Duty Paid	

All of the terms require a named place, port or destination to be identified. For example, the last term DDP is, in effect, delivered to a 'named site'. However, it is not intended to go into any further description of each of the terms here – any self-respecting purchasing department should have a copy of the Incoterms book.

Nevertheless, we should be clear about the delivery risks involved, as illustrated in Figure 5.2 overleaf.

The highest risk for the purchaser is for the supplier to deliver their goods outside the factory gates (Ex works), leaving the client to collect the goods. The risk is reduced considerably if the supplier delivers the goods onto a ship (FOB) – assuming that we are contracting internationally. Finally, the buyer's risk is at a minimum if the supplier is required to deliver the goods to the client's project location (DDP to site).

Scope of supply

The expressions listed under this heading are options for packaging and organizing contracts at the project phase level. Construction Management (Contract Management would be a better term in order to make it more generic) and Management Contracting are most commonly used in the civil and building industries. The arrangements are used to provide additional management services to the client, due to their lack of expertise or resources, to carry out the project management function themselves. In both arrangements the contractor becomes part of the client's management team, thus enabling the earlier involvement of a contractor in the development of the design. The client must, however, manage the interfaces and associated risk, but some of the execution risk has now been passed to a contractor. The primary difference between the two is that in contract/construction management, the construction contracts are placed directly with the client – in effect, a consortium of what are termed Works Contracts. In management contracting the construction contracts are placed with the management contractor, thus transferring more responsibility and risk on to the contractor. The terms are, as already stated, methods of structuring and organizing a number of contracts between the different parties in a building project, namely; client, management entity, cost consultant, architect and works contractors (there were 60 works contracts for the Scottish Parliament building.) The management and professional services' contracts are usually, and traditionally, on a structured percentage fee basis. However, there is no reason why some elements could not be on a fixed fee basis or set at a guaranteed maximum, that is, capped. The intent is that the construction management contractor formulates and places the works' contracts on a contractor managed fixed price basis.

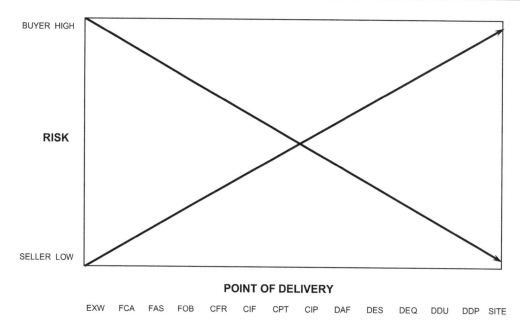

POINT OF DELIVERY

EXW FCA FAS FOB CFR CIF CPT CIP DAF DES DEQ DDU DDP SITE

Figure 5.2   Delivery risk

The antiquated nature of these methods is described in the *Official Souvenir Guide to Buckingham Palace*[10]. In discussing the transformation of Buckingham House into Buckingham Palace it states:

> *The undertaking of the building work [by Edward Blore], from 1847 to 1850, was organised on a completely different basis from the system in place under Nash twenty years earlier. Whereas then, each trade had negotiated separately with the architect for their particular component of the building, this time the entire work was put in the hands of a single contractor, who was in turn responsible for placing subcontracts with the specialist trades.*

The underlying principle of the expressions design and build, Engineering, Procurement and Construction' (EPC), EPIC and turnkey are the same, and that is the intent to eliminate the interface problems between each phase of project development. Design and Build tends to be used in building and civil engineering. When this concept was first used the industries involved got very excited, thinking that they had discovered a new concept. Whereas, in fact, it was the same as EPC that had been used in the power and process industries for many years. EPIC is used in the offshore industry and was made up deliberately by a project manager at Shell to emphasize that the offshore industry was different – an unnecessary divide – since the principles are the same. It is just significantly more expensive if a mistake is made.

The 'C' in EPC almost invariably involves some element of commissioning – varying from mechanical completion to proving performance. This is the same as the 'C' in EPIC. Thus, the only supposed difference is the Installation part – a term that ignores or leaves out the fabrication element. The lifting and installation of heavy modules (in difficult environmental conditions) is, however, only a specialized form of construction that can be performed onshore as well as offshore.

Getting the contractor to be responsible for the commissioning phase makes them focus more on the quality of their work and, thus, reduces the client's risk further.

---

10   Published by Royal Collection Enterprises Ltd, St James's Palace. Reprinted 2007, ISBN 978 1 902163 95 6.

Proving the performance of equipment can be done totally by the manufacturer or supplier. However, a wise owner will ask for their personnel to be involved in the process. This enables the owner's personnel to become familiar with the equipment that they will be responsible for maintaining.

Proving the performance of a complete facility will more often be done by the owner or client in conjunction with the contractor's, manufacturer's or supplier's personnel. The contractor may not have the operating expertise and, more importantly, the risks have changed and hazardous materials may be involved.

A turnkey contract puts all of the responsibility of proving performance onto the supplier or contractor. Although one could have a reimbursable turnkey contract, one would usually expect it to be a fixed price contract. It functions as the words imply – all the owner or client has to do is to 'turn the key or press the button' and the facility will operate. The supply of equipment is, more often than not, in effect a turnkey contract, although the term is customarily applied to the provision and proving the performance of a complete facility.

Duration of supply

The two expressions Term Contract and Concession take the concept of the turnkey contract one stage further. They extend the operation past proving performance into the operation phase for a stated period of time. For example, one might let a term contract for the provision of IT facilities or for the operation of a canteen for a period of 3 or 5 years.

A concession is similar but would involve longer timeframes. A concession is likely to involve operating a facility until the contractor has recovered their initial investment in the project. The expression would be used in conjunction with BOOT and DBFO arrangements described below.

CONTRACTOR RISK CONTRACTS

BOO and BOOT arrangements (see definitions below) are expressions describing the performance of the work. They are options for packaging and organizing work at a strategic level. They involve a concession granted by an owner to a developer, promoter or concessionaire who puts together a consortium of very different organizations to deliver the required facility. Conventional contracts will exist between the consortium members and the promoter – for example, a fixed price contract for the construction part of the work. They apply principally to the provision of infrastructure projects and utility projects – mainly power generation.

They are the only true contractor risk contracts since they involve the promoter acting as a main contractor, taking all of the risks and recouping their costs from the revenue stream. The key to setting up these contracting arrangements is in the payment terms. If a particular aspect of a facility does not meet the specified requirements, then the contractor does not get paid for that portion.

The expressions illustrate the variety of different ways that a facility can be purchased and contracts packaged together to achieve the client or owner's objectives. They also illustrate the desire that different clients have to brand their project with a different name. They are all basically BOO projects that involve a concession. In the case of a BOO project the concession is in perpetuity – for example, the Channel Tunnel. A BOOT project, on the other hand, involves transferring the facility back to the public sector after the expiry of the concession, typically 30 years – for example the Queen Elizabeth II bridge at Dartford (in this case the concession was amended so that the transfer occurred after 10 years, once the contractor had recovered their investment). Both BOO and BOOT contracts can be single contractor and owner contract arrangements, whereas DBFO (see definition below) contract arrangements will involve at least three different organizations.

Whilst BOO is commonly used in the public sector infrastructure environment, it can also apply to any owner organization developing their own facility. Upon explaining the latest fashion in contracting to one of my oldest clients they exclaimed, 'But that is how we have always done our work.' They were correct. They employed a contractor to design and construct a plant, and owned and operated it themselves in order to reap the benefits of the revenue stream. In telecommunications they use the term DAC (Design, Acquire land, Construct mast) and the operation element – in effect the maintenance of the mast – is implied. BOO and BOOT are not new but are being used more widely and more frequently.

BOO	(Build, Own, Operate)
BOL	(Build, Operate, Lease)
BOD	(Build, Operate, Deliver)
DAC	(Design, Acquire, Construct)
DBOM	(Design, Build, Operate, Maintain)
BOOM	(Build, Own, Operate, Maintain)
DBFO	(Design, Build, Finance, Operate)
DCMF	(Design, Construct, Manage, Finance)
BOOT	(Build, Own, Operate, Transfer)
BRT	(Build, Rent, Transfer)
BOT	(Build, Operate, Transfer)
BTO	(Build, Transfer, Operate)
DBOT	(Design, Build, Operate, Transfer)
BOOST	(Build, Own, Operate, Subsidize, Transfer)
FBOOT	(Finance, Build, Own, Operate, Transfer)
GOCO	(Government Owned Contractor Operated)
COCO	(Contractor Owned Contractor Operated)

GOCO and COCO tend to be used in the defence sectors. COCO is, in effect, a BOO project, and DBFO and DCMF can also have a transfer element.

All of these combinations are used in what is known generically as the Private Finance Initiative (PFI). PPP (Public Private Partnership) is similar and came into being as a result of the political protests at the time of the first Scottish Parliament elections. The Scottish Nationalists asked, 'How can you give away a public asset to the private sector?' A 30-year lease is hardly significant in the life span of a country. However, PPP gives government agencies an opportunity to take a role in a PFI arrangement – thus reaping some of the rewards but, reducing the risk transfer to the supplier or contractor.

The intent of PFI is to transfer all of the project and contracting risks to the contractor. Whilst PFI arrangements maximize the risk transfer and are at the extreme end of the contracting spectrum, some risks can still remain with the client. A good example of this was the Passport Office – 'Passport fiasco costs £12.5m in overtime and umbrellas.'[11] The reason for the 565 000 backlog of unprocessed passport applications was, as Ann Widdecombe pointed out in a question to the British parliament, 'Was it necessary to have new computer systems and new rules and regulations at the same time?' An example of two extreme risk categories being a recipe for failure. Nowhere in the press was there a mention of the contractor responsible for the development and provision of the system. The operational consequences and public relations risk remained with the client.

---

11   *The Times*, October 27, 1999.

The nature of a BOOT/PFI (single contractor) contracting arrangement means that the contractor must get the cost, time and quality targets right for the build part. If the contractor owns the project they had better get the maintenance characteristics correct and, if the contractor is to operate the facility, they had better get the performance characteristics correct. Finally, if the contractor is to transfer the facility after a specific period of time they had better get the financial model accurate. In effect, the contractor needs to do all of those things that they have always promised but never achieved.

However, with, say, a multi-organization DBFO or any other PFI arrangement, the party's objectives begin to diverge due to the long timescales involved. A contracting organization has got involved because they want the build contract. They want to get paid and then get out, leaving the operator with any problems. This means that if there is a simpler and cheaper solution it will be used, even though it may mean more maintenance work or higher operating costs in the long term.

In the 1980s when PFI was starting to be fashionable it might have taken 3 to 5 years to get a project off the ground. Government seemed to think that it could cover every eventuality and replace relationships with a contract document. In the 2000s the process, for some projects, can be finalized in approximately 1 to 2 years. Despite this improvement it is still 1 or 2 years delayed revenue.

Consequently, a contracting arrangement is needed where the schedule benefits of a reimbursable form of contract can be achieved without any negative behavioural elements.

ALLIANCES/PARTNERING/JOINT VENTURES

Whilst alliances, partnerships and joint ventures were not new, it was the CRINE (Cost Reduction In a New Era) initiative[12] that made a difference and demonstrated that they could be made to work to the benefit of all parties to a project. Alliances and partnerships are not types of contract but a relationship between contracting parties with aligned objectives.

The CRINE initiative made owners realize that they needed to take a big risk and trust the contractors and suppliers! Since it is the front-end project development phases where the real cost decisions and savings are made, the contractor, and if necessary a fabricator and or supplier, needs to be involved earlier in the owner's business and project development process. Further, by working at the relationships the risks are reduced, see Figure 5.3 overleaf.

It is interesting to note that, whilst the risk element has reduced owing to the increased focus on the relationship element, the rules element has also increased.

When I read the CRINE standard form of contract I was surprised at the biased nature of the wording; it was all in favour of the owner client. The reasoning is simple. The client can trust the contractor and supplier and, in effect, put the contract document in the bottom desk drawer and focus their attention on working together for mutual benefit. However, if it does not work, then the client needs a cast iron contract to protect themselves.

'Partnering in Europe[13] – Incentive based alliancing for projects' is an excellent report produced by the European Construction Institute (ECI) in 2001. Partnering is defined as:

Partnering is a relationship between two or more companies or organisations which is formed with the express intent of improving performance in the delivery of projects.

---

12   Launched in 1992/1993, the Cultural Change Work Group was composed of representatives from a broad cross-section of industry comprising; owners, contractors and an equipment manufacturer.

13   See *Additional Sources and Contact Details*.

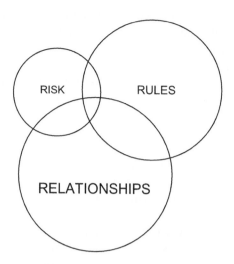

Figure 5.3   CRINE, risk rules relationships

Part 1 begins with a broader definition of Partnering and a description of two categories of the concept. Alliancing is distinguished as a specific form of partnering within one of these categories (project-specific partnering). The book states that, 'Typically, partnering falls into two broad categories:

- *Long-term partnering* (sometimes referred to as strategic alliancing). This type of arrangement generally covers the provision of services over a specific period of years and is most commonly between the owner and a single contractor.
- *Project specific partnering.* This type of arrangement lasts for the duration of a single project. In this instance, the arrangement can be between the owner and a single contractor, but more commonly it is between the owner and several contractors.'

'A variation of a project-specific partnering is where the arrangement is underpinned by an incentive scheme, whereby the rewards of the contractors and, indeed, the owner, are linked directly to actual performance during the execution phase of the project. This book deals specifically with this form of project-specific partnering, which is commonly known as alliancing.'

Addressing the relationships has been shown to be of benefit to all parties: contractors, fabricators and manufacturers. The rewards are shared on the basis of the balance between the risks. The alliance approach takes the view that it was wrong to abandon Target Cost Contracts. The rewards need to be made sufficiently attractive to align the contractor's objectives to those of the client. Under these circumstances the target cost contract works.

The client has driven the expansion of the relationship element further, to include trusting manufacturers through the use of performance specifications. However, for the supply of, say, equipment packages at a lower level of the alliance, fixed price contracts may be used. Under these circumstances one has to question if they are really part of the alliance. The nature of the contractor risk, fixed price contract discourages getting involved in examining alternative design proposals. Working closely with the other parties to the alliance will only cost them money and generate conflict over arguments about whether discussions are changes or just development work. The contractor-supplier will want to get on, produce their product and get out. There is, however, an opportunity to get involved and share in the rewards of the alliance.

A similar issue arises in the contractor's own approach to subcontracting in non-alliance projects. If alliancing works at the owner or client level why should it not work at the contractor-subcontractor level? Perhaps because contractors, being risk averse, tend to want to cascade down the risks in their main contract to the subcontracts. Further, once the project has progressed to the stage of enquiries for subcontracts the definition is such that fixed price subcontracts can be awarded. This means that the management and supervision of subcontracts can be reduced with consequent cost savings to the contractor. However, if they also learnt to use a reimbursable alliance approach with their subcontractors, might they not reap greater rewards?

'Part 2 [of Partnering in Europe] is the main part of the book. It is presented as a toolkit or "road map" for those wishing to set up an alliance. Each of the sections deals with a specific aspect of creating a successful alliance. It starts by giving an overview of the alliance implementation process, and then proceeds to cover the process of achieving commitment and alignment within the owner's organization. This is followed by guidance on the selection of alliance contractors and a short section that gives some guidance on how owners approach starting the process with potential alliance contractors.'

Prior to the publication of this ECI book my understanding of the terms was the opposite of the way that they use them. I was under the impression that the term partnering was not liked by some organizations due to the legal implications of the word. This goes to show how little common agreement exists on the terminology of contracting. As indicated in the last chapter, one should always ask the other party that one is dealing with, what their understanding is of the various contract names.

A Joint Venture (also known as a JV) is not a separate type of contract; it is a relationship between two or more parties. It is, to all intents and purposes, an alliance that has been set up as separate company from the parent companies and as a result they can be paper organizations, that is, they have no assets or substance. The key criteria for a true joint venture is that the parties have 'joint and several liabilities'. So that if one party defaults the client can claim against the other party to the joint venture. It is, therefore, important that the purchaser has recourse to parent company guarantees to be able to make meaningful claims against a joint venture.

It is surprising that it was surprising to Lord Fraser, 'That the architectural commission for the new Scottish Parliament building was let to a company with a nominal share capital of £100 and only £2 issued and fully paid up' and initially no parent company guarantee was provided!

A joint approach may be made by contractors in a tendering situation. On one occasion I was walking down the office corridor and bumped into a past boss. Since he was the UK representative of a major French contractor we were competing against on a current proposal, I naturally challenged him with, 'What are you doing here?' His response was, 'Well we are both sufficiently arrogant to think that we will each win the contract. Fortunately, someone has realized that both of us will not win and suggested that 50 per cent is better than gambling on all or nothing.' We submitted a joint proposal and the joint venture was awarded the contract. Clients need to be aware of this strategy for reducing or eliminating the competition on a tender list.

Alternatively, after having evaluated the tenders, a client might suggest to two contractors to consider a joint approach. Contractor A might have a good idea, but contractor B may be better at project management.

One of the first alliances in the North Sea was not necessarily seen as a success by the client concerned. It generated such large savings for the contractors and suppliers involved that their expectations, from the client's perspective, were set at too high a level.

Digressing for a moment to explain briefly what happens in the car components industry: the good news is that you, the supplier, have been awarded the contract. The bad news is, however, that next year, since you will have learnt how to work with us as the client, we will want the prices reduced by, say, 10 per cent. Further, in year three we will want another reduction in the price since you will have found new ways of reducing costs. In year four you may have run out of resourceful ideas or, alternatively, the relationship may be perceived as getting too complacent. Consequently, the client will invite a new round of competitive tenders in order to generate innovation.

Whilst it is not suggested that the offshore industry borrowed directly from the car components industry, this is, to all intents and purposes, what happened. On the next and subsequent contracts the target cost hurdle was lowered. A perfectly sensible step to take, however it is perceived by the contractors and suppliers as the client taking advantage of them, and they then question the merits of an alliance. The result has been that relationships in the North Sea industry in 2004/5 deteriorated and many contractors' personnel said that they preferred to return to conventional contracting.

6 *Contract Law*

This chapter is primarily about the legal aspects of contracts between individuals, or companies as individuals, in the context of English Civil Law. The English legal system, in order to be consistent, is based on the principle that a decision made by a senior court is binding on a more junior court in subsequent cases, where the facts are similar. This is known as judicial precedent, or more simply, as precedent. The law developed in this manner is often termed Common Law. Other sources of law are: legislation, regulations and directives of the European Union.

This chapter will cover the basic issues that a project manager or buyer must know if they are buying goods and services. *Ignorantia juris haud excusant* – ignorance of the law is no excuse. Nevertheless, it is not intended to be a comprehensive rendering of contract law. The intention is to provide sufficient understanding of the subject to enable the project manager and buyer to carry out their normal business and talk to their company lawyer in a common language.

Smart project managers should not try to be clever with contracts. They should be clever at making commercial and risk decisions.

Much of what follows should convey a project manager's perspective, but listening to my friend, colleague and legal guru David Wright, has also heavily influenced it.

THE MEANING OF WORDS

It is very common for people to feel that there is a barrier between their own particular area of expertise and the, so-called, mysteries of the law. This is, perhaps, because the purer the technologies the less comfortable people are at manipulating words. This reluctance to tackle the law can disappear once one has taken on board that it is simply about the meaning of words. Lawyers are good at understanding, using and manipulating words. They interpret what the words of the contract mean in the circumstances that the parties find themselves. They use rules as guidance and the first rule is the literal meaning of the words. In my early, more naive, days I thought that words meant what I intended.

To quote 'Alice Through the Looking-Glass':

'If I'd meant that, I'd have said it,' said Humpty Dumpty.

'When I use a word,' Humpty Dumpty said, in a rather scornful tone, 'it means just what I choose it to mean – neither more nor less.'

'The question is,' said Alice, 'whether you can make words mean so many different things.'

In any contract for a complex project there are an awful lot of detailed words. As a result it is all too easy to make a mistake in the language and to end up saying something that you did not mean to say, or to fail to understand what the contract actually does say. Your company lawyer should be able to give you a more correct interpretation of what the words mean, in the particular circumstances of the contract you are managing. If they are uncertain of their interpretation, you can find out (at some expense) by consulting legal counsel who specialize in your particular circumstances. If you want to pursue your point of view further you can find out if you are right (at considerable expense) by going to court. As Mr Justice Darling said, 'The law courts of England are open to all men like the doors of the Ritz Hotel.'

A judge will decide what the words mean after listening to your barrister's presentation of all the past cases supporting their argument, and the other side's barrister who also has a long list of precedents to support their interpretation of the contract words. The judge makes their decision (guided by rules of interpretation) based on what a 'reasonable person' would think the words meant in the particular circumstances. In the law of contract a reasonable person is defined as: a man of normal intelligence and understanding who comprehends the business and technology in which the contract takes place, in other words, a professional person.

A brief personal experience illustrates the difficulty of defining the meaning of the word 'reasonable'. I was responsible for a contract in which the company I worked for provided all the management positions in an Egyptian company apart from the chief executive. The contract stated that the Egyptian company would provide 'reasonable' accommodation. Well, my interpretation of reasonable was different to the chief executive's interpretation of reasonable. My interpretation of reasonable was also different to the British ex-patriots interpretation of reasonable which, not surprisingly, was also different to the American ex-patriots interpretation of reasonable. It took 2 months of surveying houses and flats in Cairo to come to a common agreement of the meaning of the word 'reasonable'.

Obviously the simplest, and cheapest, method of dispute resolution is by negotiation in order to arrive at a compromise solution to the parties' differences.

CONTENTS OF A CONTRACT

The reluctance to get involved with understanding, influencing or the development of the terms of the contract often comes about due to the historical culture of an organization. I have frequently heard, 'Oh, the contracts department deal with that.' This is very common in government organizations or organizations that have their origins in government. In these situations one has to question who is managing the contract – a contracts department or the project manager.

When asking training course delegates to identify the elements that make up a contract, experience shows that almost all of the issues (initially identified) are detailed clauses forming part of the contract terms. Delegates do not seem to understand that the other elements, listed below, are just as much contractual as the small print of the contract:

- the payment terms;
- the scope of services;
- the scope of work;
- the specification;
- the design basis;
- standards;
- the programme;
- the organization;
- the administrative requirements.

Strictly speaking, the payment terms are part of the terms of the contract. However, their commercial content is sufficiently different for them to be identified separately, and may be the responsibility of a commercial department and the project manager. Similarly, the project manager may well have compiled the scope of services and scope of work in conjunction with an operations department. The specification and standards will be contributed by the technical departments, and project controls personnel will coordinate all departments' requirements to produce an outline schedule. In effect, the project management team is probably responsible for more input to the contract than the legal or contracts department. The intent is not to disparage the importance of a legal department. They are a vital element in the organizations risk management processes, and they are there to help us avoid problems created through the use of incorrect words. They are specialists like other technical departments. As JP Morgan said, 'I employ a lawyer to help me do what I want to do. Not to tell me what I can't do.' Nevertheless, the lawyer, as a key member of the risk management team, should tell people when they are being too clever and creative and are crossing the legal boundaries.

Even within the terms of the contract there are clauses where the project manager should have significant influence, if not, input. For example:

- change order clause;
- liabilities and liquidated damages;
- allocation of risks;
- *force majeure*;
- project completion and hand over.

FORMATION OF A CONTRACT

To enable business to take place in an international marketplace, contract law has to be fairly straightforward. It has to be if it is to cope with situations as diverse as buying goods across a counter and designing and constructing a process plant. It can be summarized as follows:

You may make any bargain you choose (provided it does not conflict with Statute law, that is, the national interest). However, having made the bargain you must fulfil your promise or compensate the other party for the loss that they have incurred.

In effect and in general, the attitude of the law is, 'Make your contractual bed then lie on it.' As such, it should be noted that the law regards business as being fully conversant with, and responsible for, any obligations or agreements contained in the contract regardless of their fairness. Whilst there are a few National Statutes such as the Unfair Contract Terms Act, these are very much backstops that aim to prevent the very worst of business abuses.

Let us begin by examining the start of the contracting process. In a project context the process usually starts with a client issuing an enquiry. It is, of course, possible for a contractor to initiate the process by suggesting that they could provide goods and services to a prospective client. However, an enquiry (or an approach of the nature suggested) is only an Invitation To Treat. An invitation to treat can be goods displayed with prices in a shop, an advertisement, a catalogue listing equipment and materials, and an enquiry document for goods and/or services. However, the invitation to treat is not part of the contract (see later).

There are three fundamentals for a contract to be effected:

1. offer
2. acceptance
3. consideration.

There are some other tests to establish that it is a valid contract, but these are the essentials. The other elements are:

4.   intention to create legal relations;
5.   capacity;
6.   legal and possible;
7.   form;
8.   certainty of terms.

In general, a contract will exist when an agreement is made by acceptance (without qualification) of an offer. The promise (to supply goods and or perform services) should be made with the intention that it should be binding when accepted, and the acceptance must be communicated to the party making the offer. The agreement must be a bargain (not necessarily equal) and not mere promises. There must be some consideration that passes from the person accepting the offer to the person promising to supply goods or services.

Offer and acceptance

If company X offers to supply a certain item for £1000 and company Y accepts this price, a contract is created. If, however, X advertises the item for £1000 this is not an offer but an invitation to treat. An invitation to treat must be distinguished from sales and marketing language, and an offer must be distinguished from simply the transmission of information. This means that Y would have to offer the full £1000 and X would then have to communicate acceptance of the offer in order for a contract to be created. If Y only offers £950 then this is what is known as a counter offer, which would then have to be accepted by X before a contract could be established. A counter offer is a rejection of the original offer and the original offer is no longer available for acceptance.

This may sound very pedantic but it is often the case in business that a company will say something like, 'We accept the contract subject to the following...' This is a counter offer which the first party must unconditionally accept before a contract is created. On the other hand, a question asking whether the other party might accept certain changes to a specification is not a rejection, but a request for information.

Acceptance 'subject to contract' does not form a contract. It is neither acceptance nor rejection. It implies that the parties are agreed on the terms but neither party is bound until a formal contract is signed. However, 'agreement to agree' achieves nothing, since the parties may never agree.

If a transaction is made without the final acceptance, and if things go wrong, the existence of a contract may have to be decided by a court of law. In this case the courts will carefully consider the actions of both parties, their previous dealings with each other and any correspondence entered into before deciding who is liable for what. For example, starting work on a project may create a contract by performance, if that has been the custom and practice between the parties.

Communication of acceptance

In general, acceptance must be communicated to the other party. 'Unless we hear from you by Friday we shall assume that you have agreed...' does not form a contract. Silence is not acceptance – there must be some action to indicate acceptance. The party making the offer may request that acceptance is in a particular form, for example, by telephone, in writing or by letter.

If acceptance is by letter then the contract is formed when the letter is posted. Whereas, a faxed acceptance will only create the contract when the fax is received.

Intention

It must be in the minds of both parties when the contract is made that a contractual relationship is being created. At one extreme the law ignores social, domestic or family contracts, such as dinner parties. However, in business the law assumes a high level of competence on the part of both parties. It should be noted that one party cannot claim later that they did not realize that a contract was being created if, by their actions, it is obvious that they did.

Furthermore, companies have to be very careful to define clearly, to those with whom they do business, who in the organization is permitted to enter into contractual commitments on behalf of the organization. If a buyer was reasonably led to believe that the person with whom they were dealing had permission to act on behalf of their company, and the company took no action to dispel this assumption, then the contract can be binding on the company. This would be the case even if the company then claimed that the person in question acted outside their limits of authority.

Consideration

Consideration is the legal expression used to define what has to be given in return for the supply of goods or services, in order to create a bargain. Money is only one example of consideration – legally it has been defined classically as, 'Some right, interest, profit or benefit accruing to one party, or some forbearance, detriment, loss or responsibility given, suffered or undertaken by the other.' That is, it must have some value to the recipient.

It should be noted that the law does not exist to repair bad bargains. It assumes that the parties involved are capable of looking after their own interests and as such the amount involved is irrelevant. One sometimes hears about goods or services changing hands for nominal sums. For example, the Government sold Concorde for 5 pence – my parents received a peppercorn rent for an aerial wire that traversed the roof of their house in London. This is not a quaint old tradition but a way of ensuring that the contract involves consideration – a contract must be a bargain not a mere promise.

Capacity

Obviously the law insists that the parties entering into the contract have capacity, that is, they are not very small children, certified insane or certified alcoholics. This means that most adult citizens in the UK have the capacity to contract! Whether they have the authority to do so, from their employers, is another matter (see intention).

Legal and possible

Any contract has to be legal in its intent. One cannot sue a hit man for a failed contract or a bank robber for a poor return on a raid. In addition to legality, the contract must be possible, that is, regarded as such by any reasonable person. As such, it is not possible for a person to enter into a legally binding contract to invent a perpetual motion machine or a 100 per cent efficient petrol engine.

Form

All contracts must have a form. Property, for example, must be transferred in a document 'under seal' – although an actual seal is no longer required these days. Usually contracts will be written. However this is not to say that they must always be written down. Most contracts between individuals are undertaken on a verbal basis (or in silence), as in buying petrol or a bottle of wine, for example. 'A verbal contract isn't worth the paper it isn't written on!' due to the difficulty of proof. It is interesting that, in recent years, the courts have upheld some verbal contracts. However, in business the terms associated with the transaction are usually so involved that a written format is used so as to ensure that there is less chance of misunderstanding later.

TERMS AND CONDITIONS

This statement is in itself misleading as it confuses the hierarchy of obligations contained within the contract. There are two types of terms within a contract:

Express terms

These are terms that are specifically mentioned or agreed verbally as a part of the contract. There are three types of Express Terms – Conditions, Warranties and Innominate Terms. Conditions are terms that go to 'the root of the contract'. A condition is an obligation that is so essential that its non-performance may be considered by the other party as failure to perform the contract at all. Warranties (not to be confused with guarantees), however, are subsidiary to the main purpose of the contract, and are not so vital that a failure to perform them invalidates the whole contract. For example, 'The contractor will supply lifting equipment model X.' On the way to the client's premises the equipment is involved in a road accident, so that the contractor supplies equipment model Y. The provision of lifting equipment is a condition of the contract, whereas the model number is a warranty.

It is the courts who will decide what is a condition of a contract. So that when we write, 'The following conditions of contract shall apply,' we are incorrect, although it might help sway the judge in making their decision in the event of a dispute. The crucial difference between conditions and warranties is the remedies that can be applied in the event of a Breach of Contract - see below. An innominate term is one where one cannot be certain about the remedies available.

Implied terms

These are terms that normally arise as a result of the existence of Statute Law, which applies regardless of whether or nor the parties intended that they should. An example of this might be an exclusion clause exempting one party from any liability for death or injury due to their negligence. This is not permitted in, for instance, English law. Thus, a court, despite the fact that it was written and agreed in the contract, would throw it out. Another good example of implied terms are those inferred by the Consumer Protection Act and the Sale of Goods Act that provide consumers with a safeguard regardless of the exact wording in the formal contract.

Certainty of terms

There must be certainty of the terms of the contract for the contract to be valid. That is, there must be agreement without qualification. I remember writing a report for a QC saying, 'The contract came into being on March first...','I am not sure that's true,' he responded, 'since I have found correspondence – dated 3 months later – where they are still discussing the special conditions of the contract.' 'Ah,' I replied, 'but was there a contract by performance, since they had started work?' At this stage one might think that this is a problem for the lawyers to resolve and that project management is no longer involved. This is not the case, however. As the project management expert involved in the dispute I needed to know when the contract came into being, since this affected the project start and finish dates for the programme of work and, hence, how late the project completion was.

This is a subject that is probably better handled in contracts for services but is commonly unresolved in the purchase of goods. Each party in the offer and acceptance process sends their own standard documentation (using pale coloured paper) with their own standard conditions printed on the back – using a small typeface in faded ink! All of which is designed to discourage the uninitiated from reading the proposed terms. Further, for the amateur, they do not exactly make exciting reading. Even the professionals will not read them since they will be responding stating that their own standard terms apply. This is termed the Battle of the Forms.

Standard terms

Industry and business sectors develop standard contracts so that the contracting parties can have a reasonable degree of certainty about the intent and meaning of the terms in the contract. Standard contracts are designed to produce consistency and prevent errors and omissions. Individual companies also develop standard terms on which they are willing to do business. The terms are written so as to protect them from the risks that they are not willing to accept.

Consequently, do not alter the standard company terms without checking with your legal or contracts department. The terms will have been developed to stand alone as a coherent document. If you change one item (however simple and for whatever logical reason) you may not know the consequences it can have in another part of the contract terms.

The battle of the forms

This comes into being through the use of standard documentation and standard 'terms and conditions'. The battle of the forms develops as follows:

a) A client (owner, contractor or subcontractor) requiring goods, issues an enquiry with appropriate documentation that includes their standard contract terms – an invitation to treat.

b) A supplier issues a standard (for them) quotation form with their standard terms on the reverse of the form – an offer.

c) Discussion takes over quantities and specification details – negotiation.

d) The client then issues a standard order form with quantities, delivery instructions and so on, and with their standard terms printed on the back of the form – in effect, a counter offer since there has been no acceptance, without qualification, of the offer.

   The more sophisticated client may issue the order with instructions saying, 'Please sign and return the tear-off slip at the bottom of this order.' The tear-off slip will have words similar to, 'We acknowledge receipt and acceptance of your order in accordance with the terms stated therein.' Perhaps the new secretary or administration assistant, in the supplier's organization, will use their initiative and think, 'I won't disturb the boss with such a trivial item,' and they sign and return the acknowledgement slip. This would then be acceptance without qualification and a contract would have been created the moment the tear-off slip was posted.

e) Without the tear-off slip the supplier might issue an order acknowledgement confirming quantity, delivery details and prices in accordance with their standard terms printed on the back – a counter offer.

f) The goods are manufactured and delivered. The driver of the delivery vehicle asks the stores person or gatekeeper to sign the delivery note 'for our records'. Printed on the back are the suppliers 'standard terms'.

   The well trained stores person may say; 'I am sorry that I cannot sign the goods received note, but I can stamp it.' The official looking stamp might incorporate the words, 'Received in accordance with our standard terms'!

Two or three questions remain. Has a contract been created and under whose terms? And, does it matter? If the goods have been delivered according to the client's requirements, and if the supplier gets paid, then it may not matter. Some form of contract may have been created depending on custom and practice between the parties, and there is a likelihood that the terms of the contract are the last ones to be advised to the other party.

Of course, if there is some disagreement over the nature of the goods and the parties fall out, then they will wish that they had sorted out the terms of the contract. It is much easier to come to a mutual agreement on the terms of a contract whilst both parties have the same objective (one wants to buy and the other wants to sell), than when time has elapsed and they see the issues differently. One should remember that contracts are made amongst friends for when they fall out, or something goes wrong.

Subcontracts

The contractor client should take care not to impose an unreasonable number of lengthy terms onto their subcontractors or suppliers. Just because they have agreed onerous terms with their client, it does not mean that the terms are appropriate for the subcontract. Contracting is about understanding and taking appropriate risks – and protecting oneself accordingly. The contractor client should be able to sort through the terms that they have agreed and decide what is appropriate to pass on to the subcontractor. Alternatively they should develop standard 'short forms' of contract suitable for smaller companies. There is no point in passing on liabilities of, say £1million, if the assets of the small painting subcontractor are only £100 000.

SOME SPECIFIC ISSUES

Termination of an offer

Either party to the process can terminate the offer. One party can withdraw the offer (before acceptance), and the other party can simply reject the offer.

As has already been demonstrated, an offer is also terminated by a counter offer. If a time limit was stated in the offer (always good practice for a supplier or contractor in a tendering situation) then the offer will lapse after passing the deadline.

If an offer was made subject to a particular condition, which is not met, then the offer is terminated.

Finally, if one of the parties dies the offer dies with them. However, this cannot happen with offers made by companies.

Letter of intent

A letter of intent is sometimes entitled an Intention To Proceed (also known as an ITP). This should not be confused with an Instruction To Proceed (also known as an ITP). A letter of intent is similar to saying 'subject to contract'. That is, the formal contract is being delayed until some event has taken place. The letter will be worded so as to avoid creating any legal obligations. On the other hand, an instruction to proceed creates an obligation to pay for whatever works the supplier or contractor is asked to perform. A well-worded letter of intent will have all the various elements that are required for a contract.

Novation

Novation is the process of transferring a contract from one party to another. An owner or client may place a contract for the supply of critical materials or equipment in order to reserve manufacturing capacity. Alternatively, the client may select specific subcontractors that they wish to be used on a project before a main contract is awarded. This is common practice in the construction industry. The subcontracts are then novated to the main contract when the main contract is being finalized. In these circumstances, it is important that the enquiry document and proposed terms of contract for the main contract are structured and worded to allow for this.

Time is of the essence

Time is of the essence means that timely delivery is so important to you that failure to deliver on time, by however small an amount can be construed as a breach of contract. The danger for the supplier or contractor project manager is that project activities always seem to be late. If deliveries are running late then the purchaser can cancel the contract, purchase them elsewhere and the supplier must compensate the purchaser for the loss they have incurred. Time is of the essence is commonly used in purchase orders.

Time is of the essence is, however, a very common implied contract term in project contracts. As soon as a delivery date or programme is included in a contract it is implied that time is important to the purchaser – without specifically using the phrase 'time is of the essence'.

Force majeure

Literally this means 'major force' – that the parties have agreed would be beyond their ability to resist (or control). It is often assumed that this term has a specific meaning. Whilst this is true in French law, it is not in English law. In English law the term means just what the words written in the clause mean. Say, for example, that the clause includes the expected words of 'war, earthquake, fire, flood, strikes…' . Then a strike in a manufacturer's works might be claimed as *force majeure*. However, if the wording was, '…other than any lock-outs or other industrial disturbances relating to the company's own workforce and its subsidiaries…,' then, in this situation the strike in the manufacturer's works could not be claimed as *force majeure*, if it was a subsidiary of the company concerned.

A supplier's or contractor's project manager may see a strike as an opportunity to build up costs to back charge their client. This would be a mistake, since each party in a *force majeure* situation is expected to bear their own costs and, in addition, there is an obligation in law to mitigate one's loss (an implied term).

Independent contractor

It is sensible to have a clause stating that the contractor is an 'independent contractor'. This means that they are responsible for the method of execution, the planning, the allocation of resources and so on. Whilst this is relatively clear when dealing with the purchase of materials and equipment, it could be confused for, say, a reimbursable (client managed risk) contract. The independent contractor clause would break the contractual chain of liability and prevent a mismanaged subcontractor pursuing the client for damages.

Damages

As indicated at the beginning of this chapter, if one party to the contract fails to perform their obligations, then the other party is entitled to be compensated for the loss they have suffered, due to the failure of the bargain. The idea is to restore the injured party to the same position they would have been in if the contract had been performed.

Liquidated damages

The idea of liquidated damages is that they are pre-agreed in order to save time in arguing over the settlement of damages at the end of a contract. They also help the contracting parties to control their risk exposure. It is generally accepted that liquidated damages should be based on a genuine pre-estimate of the losses which the client would incur if the supplier or contractor fail to perform.

If the damages are for a fixed total amount (for, say, late delivery or completion), then there is a danger that the courts will view them as a penalty and reject them. Liquidated damages for a delivery delay should be time related on, say, a per week or percentage basis.

DISCHARGE OF A CONTRACT

A contract can be discharged in a number of ways: for example, by:

* Performance: both parties fulfil their obligations.
* Agreement: both parties agree to waive the contract.
* Frustration: for example, war, riot or earthquake.
* Breach: parties to the contract fail to fulfil their obligations.
* Others: death or bankruptcy.

Performance

The normal method of discharging a contract is for each party to perform their obligations required by the contract. The performance has to be complete and exact. This requirement is the client's opportunity to make the supplier or contractor 'dot the i's and cross the t's' so that the client gets what they are paying for. However, the contractor may be able to claim substantial completion and get paid for the work done, less an amount to cover the outstanding work. Under these circumstances the client loses the leverage to make the contractor perform.

Agreement

There is nothing to stop both parties agreeing to terminate the contract amicably. After all, they agreed to the original contract, so they can agree to a different contract to terminate the original one – but, there must be some consideration involved.

Breach of contract

A party will be in breach of contract if they do not perform their obligations precisely. There are a number of ways that this can occur and the most important is a breach of a contract condition. Whilst the breach of a condition gives the client the right to terminate the contract, their problem is the difficulty and inconvenience involved. However, since a breach of contract (condition or warranty) will give rise to the requirement to pay damages, it is the damages that are important.

Breach of an innominate term will also allow a claim for damages but one cannot be certain about the right to terminate the contract, except in the case of a very serious breach of contract. This is an example where you go and talk to the specialist lawyers.

There can also be circumstances where a party may anticipate being in breach of contract. In the above example of the contractor supplying lifting equipment, the contractor could advise the client that they anticipated being in breach of contract due to frustration.

Frustration

This is what it says it is. Some outside event totally prevents one of the parties from fulfilling their obligations. As a result, the contract comes to an end and monies paid have to be repaid. However, it is very rare for a commercial contract to be totally prevented.

PERFORMING THE CONTRACT

I have always believed that, as a project manager, one should be totally familiar with the detailed contents and implications of the contract. You then place it in the bottom drawer of your desk and forget about it. The next time you need to check the contract it should be to verify specific deliverables and handover details. If you need to read the contract before then your project is probably heading for trouble.

Having said this, contracts for large projects in the 2000s are very complex and it is necessary to keep checking them to ensure that contractual obligations are being met. Further, do not forget your own obligations as the client.

The above approach is very similar to the Chinese (1980s) attitude. The Chinese did not regard the contract as set in stone, but as a document that was to be interpreted on the understanding that both parties were working together to reach an agreed goal. In the course of executing the project they did not seek to exploit every advantage, nor did they expect the other party to do so. They considered that the contract was a guide, and that behaviour could be varied to meet circumstances as they arose.

However, if your behaviour is such that your relationships break down, there are other options, available rather than ending up in a court of law.

Dispute resolution

Arbitration sounds an attractive solution to settling disputes instead of ending up in the very expensive law courts. However, there are always at least two sides to a story and arbitrators too often end up with 50/50 decisions, resulting in both sides feeling dissatisfied. The result is that one party may still exercise their option to take the matter to court.

Senior management will not be enthusiastic about arbitration or going to court. The chances of, so-called, winning will be dubious and it will be expensive and time consuming. Senior management's job is to run the business and not to spend time in discussions with lawyers. Further, in order to satisfy both contracting parties need for impartiality in the arbitration process, a neutral location such as Switzerland may be chosen. From a project management perspective this is disastrous. Think of the cost of transporting all the files and the travel and living expenses of the people who will be needed for indeterminate periods of time! The project team should be managing a new project not sitting around waiting to give evidence.

Finally, if you have to end up pursuing a legal route the whole process should be treated as a project in its own right and project managed. It will be important to recognize that it will require give and take in order to reach agreement.

If you can keep your cool you will be much better off negotiating a deal that may have a better chance of satisfying both parties. To help the process the concept of Alternative Dispute Resolution (also known as ADR) is used. This simulates a legal process using an independent facilitator in order to negotiate a settlement. The mediator gets to know the real interests of the parties and, as a consequence, can identify solutions that the individual organizations would not be able to identify. Accordingly, they can guide the process so that the parties end up with their own win-win resolution.

There are many variants and subtleties to the whole of contract law and in particular, the 'specific issues' highlighted above. Consequently, anyone experiencing any of them, who does not discuss them with the company lawyer, is taking unacceptable risks. However, whenever you ask a lawyer about a particular issue concerning your project, they will always respond with, 'It depends on what the words in the contract say.' Consequently, if it is an issue that requires interpretation do not try and do it yourself. Read the contract words and then discuss it. As I believe one judge was heard to say, 'A man who is his own lawyer has a fool for a client.'

CONTRACTING EXERCISE

The following exercise simulates the contract contents and the process of forming a contract that has been discussed in this chapter. Analyze each of the numbered documents to test your understanding of the issues, determine the status of the documents and answer the following questions:

- Does a Contract exist for the helicopter trip?
    - If so (and even if you think no contract was made) what are your answers to the rest of the questions?
- What does the Scope include?
- What is the Price?
- On what Date was it made?
- Can the contract be cancelled?

FLY ME TO THE MOON (and BACK)????? by David Wright

Document 1

SALES DIRECTOR MAGAZINE (Advertisements section) 10th April

EXECUTIVE TRAVEL
FLYME HELICOPTERS Limited, Cranfield Airport, Bedford.
Executive Helicopters for hire, complete with pilot and in-flight attendant, for one-off or regular charters. Based near Bedford and Milton Keynes, our aircraft will fly you and your important customer there and back again.

Six-seater helicopter daily hire rate, complete with crew, £800 per flying-hour, (£500 per hour on the ground), for all journeys from our base at Cranfield airport to any destination in England.

Interested? To book your helicopter write, FAX or telephone our Sales Director, Charles Flyme, at the following address ....

Document 2

FAIRY FOODS LIMITED
Bedford.

FLYME HELICOPTERS Ltd   14th April
Charles Flyme, Sales Director.

Dear Mr Flyme

Your Advertisement in the Sales Director last week is of interest to us. We expect the Managing Director of an important Far East customer to visit us here in Bedford in six weeks time. He will be accompanied by his wife, who is a very keen gardener, and we know that they would like to visit the gardens at Tresco Vean in the Scillies.

We, therefore, write to place a firm booking for a six-seater helicopter for Saturday 31st May to take five people to the Scillies and return. A friend of mine tells me the flight to the Scillies will take two and a half hours, so based on the rates in your advertisement, can we agree a price of £6,000 for the return trip plus a four hour stay?

Yours sincerely

Herbert Fairy
Managing Director

Document 3

FACSIMILE
FLYME HELICOPTERS Ltd.

FAIRY FOODS Ltd   17th April

Attention: Mr H Fairy

Thank you for your letter of the fourteenth.

We have made a firm booking for your company for our six-seater helicopter covering a full-day visit to the Scillies for five passengers.

Unfortunately the flight will actually take rather longer than you have been informed, and there are special landing fees payable in the Scillies.

The price will, therefore, need to be adjusted to £6,800. I will assume that this is acceptable to you unless I hear to the contrary within the next ten days.

Do you wish Flyme to provide in-flight drinks and hot meals? (Light refreshments are already included in the price.) If so, our charge would be £40 per person.

Regards

Martin Flyme
Commercial Manager

Document 4

FACSIMILE
FAIRY FOODS LIMITED

FLYME HELICOPTERS Ltd   6th May

Attention: Charles Flyme – Sales Director

Dear Charles

So nice to meet you at the Old Boys dinner last week. I had never realized that we had actually played in the same cricket team together.

  We accept the terms set out in Martin's fax of 17th April, and could you also add drinks and meals for five, making a round price of £7,000. However, we have a change of plan. For business reasons we need to switch the destination to Quimper in Brittany.

Regards

Bill Fairy
Sales Director

Document 5

FACSIMILE
FLYME HELICOPTERS Ltd.

FAIRY FOODS Ltd   8th May

Attention: Mr W Fairy

Thank you for your fax earlier this week.

  We have changed the booking as you requested. However, there is a landing fee of £100 payable in Brittany – so our costs will increase accordingly.

Regards

Martin Flyme

Document 6

FACSIMILE
FAIRY FOODS LIMITED

FLYME HELICOPTERS Ltd   12th May

Attention: Martin Flyme – Commercial Manager

Dear Martin
Looking at this whole trip the price has gone up from around £6,000 to over £7,000. I hate to say it but it's beginning to look rather expensive.

Could we cut out the refreshments to keep the price below £7,000? I understand that the accountants will be happier if we can do that.

Herbert Fairy
MD

Document 7

(Sent by facsimile at 10.45 hours)

FACSIMILE
FLYME HELICOPTERS Ltd.
Cranfield.

FLYME HELICOPTERS Ltd   15th May
Attention: Mr H Fairy
Managing Director

Dear Mr Fairy
Sorry if the money men are giving you a hard time. Actually I think I CAN do something rather special for you.

If you would agree to switch to a five-seater, and a short stop in Southampton on the outward journey, of perhaps 15 minutes, to allow the crew to off-load a package that we have to deliver for another customer. If so we could cut our price to below £6,000 for the complete package.

Yours sincerely,

Charles Flyme
Sales Director

Document 8

FAIRY FOODS LIMITED
Bedford.

FLYME HELICOPTERS Ltd   15th April

Attention: Charles Flyme.

Sales Director.

Dear Charles

As you may have heard my father was taken ill a week ago, a rather bad bout of influenza unfortunately, and I have been away on business for the last week.

I have just remembered that I have not yet written to you to confirm our arrangements for the 31st.

Please accept this letter as confirming that we agree the terms of FLYME HELICOPTERS last fax.

Yours

Bill Fairy

Document 9

KOBE FROZEN FOODS kkk
Kyoto

FACSIMILE   Saturday 24th May

To: Mr H Fairy

URGENT

Concerning planned visit to Bedford. Eldest daughter has just produced first grandson, 3 weeks early. Both well and congratulations very much in order, but wife now refuses to travel anywhere for one month, and I also. Much regret therefore but must postpone tomorrow's visit for six weeks. Many apologies for any trouble caused.

Regards

Discussion and Analysis

This exercise sets out a typical exchange of correspondence between two organizations both fictional and, it is contended, real life. It illustrates the need for clarity in the communication process and the requirement for one person to be in charge.

Document 1

This is an Invitation To Treat to the world at large – there is an intention to create legal relations. It involves a legal activity (it is not intended for criminal purposes) and the activity is physically possible. As a limited liability company it is safe to assume that they intend to trade, so they have the legal capacity. It describes the scope of service available and identifies the price (consideration).

Document 2

As requested in the invitation to treat this letter is addressed to the sales director. The first paragraph of the letter describes the scope (destination) of the service required. The second paragraph makes an offer identifying a programme (date) and offers consideration. There is clearly an intention to create legal relations.

It appears that a contract is about to be made. There is an invitation to treat and a clear offer. All that is needed to finalize the contract is a clear acceptance. However, as is common in commercial transactions, a counter offer (in Document 3) confuses the picture.

Document 3

Paragraphs three and four of this fax make a counter offer for the issues raised so far. Paragraph five increases the scope of the service and makes an offer in relation to the additional services.

Sales, having done their job, pass the fax on to the commercial department to finalize details. However, this fax is, in effect, a counter offer and all previous offers die with it.

In a further attempt to finalize a deal the commercial manager says, 'I assume that this is acceptable to you unless I hear to the contrary within the ten days.'

Document 4

This document is dated 6th May – more than 10 days after document three. Consequently, one might think that a contract has been created since no statement has been made rejecting the counter offer in document three. Silence does not imply acceptance.

The second paragraph, however, appears to accept Flyme's offer in document 3 but then makes a counter offer of £7000. It further confuses the matter by making a scope change for a different location. This fax is another counter offer.

It should be noted that the MD of Fairy Foods has passed the responsibility for finalizing the details of the proposed trip to his sales director. Nothing unusual in that but Fairy's sales director has written to his opposite number, Flyme's sales director – not the commercial manager. Are the sales and commercial people keeping each other up to date with the latest information? However, there is clearly a strong relationship between the parties as indicated in paragraph one and the reference to 'Martin' in paragraph two.

Document 5

This fax could be a counter offer but there is a lack of clarity regarding the consideration. Whilst a price to one party becomes a cost to another party, we cannot be certain that Flyme's increase in costs is intended to be an increase to their price. Is it just a statement of fact? This is an example of the importance of using the correct words.

Document 6

The MD of Fairy Foods is clearly unhappy (but still on first name terms) and is taking control, but his phraseology is confusing. Could the first paragraph be a cancellation of the counter offer of 8 May (document 5)? Further, is paragraph two a question or a counter offer?

It is interesting that he is blaming the accountants, rather than admitting that he is the one who is unhappy!

Document 7

Martin, Flyme's commercial manager, has upset the potential client and so sales have taken over, despite not having such a close relationship (Dear Mr rather than first names). Sales will save the deal with 'something rather special'. A counter offer cutting the price below £6000. However, it is unclear how much below. Consequently, there is uncertainty regarding the terms for the consideration.

Not such a special deal considering that they will probably be charging their other client (delivery of a package) for the trip to Southampton. So approximately 40 per cent of Fairy's trip will be paid for and Flyme only offers a £1000 reduction! Since Fairy Foods do not appear to have a commercial manager they may not realize that they could negotiate a better deal.

Document 8

This is where matters get even more confusing. Firstly, what is the date of this letter? It cannot be the 15th of April since the original offer (document 2) was dated 14th April. Further, it refers to 'Martin's Fax of the 8th of this month' (document 5). Clearly this is an obvious error and a judge would agree that the letter should be dated 15th May.

On the face of it, paragraph three, the last paragraph, looks like a clear acceptance – but of what? Bill Fairy (under emotional and business pressure) has not read the file properly. Did Herbert Fairy's fax of the 12th May (document 6) cancel the trip? If the last paragraph of document 6 is a counter offer (rather than a question) it would supersede all previous offers. Further, the document 7 letter is an offer from Flyme superseding the offer in document 5 – but, when was it received by Fairy Foods?

An offer made by letter is valid as soon as it is posted. However, this letter was faxed, and a fax is only valid when it is received. Let us make the reasonable assumption that both parties dictated their letters first thing in the morning. Bill Fairy may well have said, 'Get it in the post straight away.' On the other hand, Charles Flyme, having dictated the letter, had it typed up and having signed it said, 'You had better send it by fax.' So, was document 8 posted before document 7 was received, or vice versa? If document 8 is a valid acceptance of document 5 then Bill Fairy has accepted a deal that is more expensive than his father (Herbert Fairy) wanted, and has missed the opportunity for a better deal in document 7. However, the consideration or price in document 5 is unclear. Therefore, there is uncertainty of terms and a contract would not exist.

If, on the other hand, document 8 were an acceptance of document 7, a contract would exist on terms that would make Herbert Fairy really pleased with the commercial skills of his son.

Document 9

Clearly, the cancellation of Kobe's visit will result in the cancellation of Fairy Food's contract (if one exists) with Flyme Helicopters. This is a fax from a third party that is not a party to the (possible) contract. The concept of 'privity of contract' says that a contract cannot confer rights or impose obligations on any person who is not a party to the contract. Kobe Foods are, therefore, within their rights to cancel their arrangements with Fairy Foods without any consequences (for them) arising out of the cancellation of Fairy's contract with Flyme.

Charles Flyme will most likely believe that a contract exists and may be upset at the cancellation and could claim damages, from Fairy Foods, for their failure to fulfil their contract. The result would be a loss of the relationship at a considerable cost paid to both parties' lawyers. A better solution would be for Flyme to negotiate a deal for the trip when it is reinstated.

Conclusion

The conclusion for Question 1 – whether or not a contract exists, is somewhat uncertain.

Question 2 concerning the scope is a little clearer: a five-seater helicopter to go to Quimper in Brittany with drinks and meals. Whether or not the trip goes via Southampton depends upon the status of documents 5 and 7. However, in reality it is likely to include Southampton and, of course, the return trip.

Question 3 regarding the price is less clear. Is it £7000, or £7000 plus £100 extra cost (document 5), or is it below £6000 (document 7)?

Question 4: If a contract was made it looks as if it was made on 15th May.

Question 5: The only easy question with a right answer. If there was a contract the answer is, yes it can be cancelled (see brief discussion under document 9 above).

Key learning points from this exercise:

- one person to be in charge of the purchasing process;
- understand the basic rules for creating a contract;
- the importance of clarity in the 'offer and acceptance' process;
- clarity and consistency in the communication process;
- proper administration (filing, numbering and circulation) of the documentation;
- agree the scope first and then the price follows;
- understand the meaning of key words, and the correct use of words;
- regardless of the Rules, the Relationship may be more important.

7 Communicating the Requirements

I was on a sailing cruise down the coast of Thailand one day before the Tsunami struck and met a past President of the Institution of Purchasing and Supply. As he was currently head of procurement for a major corporation I realized that I was talking to an expert. So I asked him, 'What is the secret to purchasing?' and he replied, 'I still think the secret to purchasing is getting the need right.' He was right, there is no point in negotiating a superb deal for 10 000 units of office supplies if you only need 5000 units because they have just cost you twice as much. The same is true in a project environment, the secret is getting the scope right. Whilst we think we can do this reasonably well, one thing project people find incredibly difficult to get right is the quantities for materials. We always seem to be ordering half a dozen more. *Time spent getting the brief, the scope, the specification or the needs right, is never wasted.*

The subject of identifying what is required is not a buyer's function. However, it is an integral part of the procurement procedure and defining what is wanted is one of the most important aspects of the process. Having said this, at the start of the buying process, people do not really know precisely what they want or how to define it. Consequently, a buyer should be aware of the technical communication process and what to look for in a specification. The buyer should review all of the enquiry documentation for consistency, clarity of language and avoidance of jargon.

Many surveys have shown that the common cause for project failure is failure to define what one wants. There are two aspects to the definition process. Firstly, the Scope of the product or services to be supplied and, secondly, the details of the features of the product or services – the Specification. In this context the term scope is used to describe the boundaries of the macro elements – the deliverables that are required. The term specification is used to describe the micro-level details or performance aspects of what is being purchased. Conversely, a specification can also be a document to describe what a tenderer or vendor is willing to supply to a purchaser.

For simple requirements a two- or three-page letter may be sufficient. For a complex facility involving equipment, materials and services it may take a hundred pages or more to define the requirements. Both of these approaches have deficiencies. The first may be incomplete, have omissions or lack sufficient detail, resulting in an unsatisfactory tender. The second may put off the tenderer due to the work involved. Consequently, it is necessary to develop mechanisms that communicate the requirements in as succinct a manner as possible.

Organizations with well developed procurement processes have very prescribed mechanisms for ensuring clarity of the communication process between buyer and seller. In these organizations the specification will form part of a Material Requisition. The purpose of the material requisition is to convey information concerning material, equipment and/or services from the design department, or the person responsible for the technical aspects of a project, to the purchasing department. The material requisition/specification/technical specification is a document that uses words, drawings and data to define what you, the purchaser, wants. *These documents form the technical terms of the proposed contract.*

Requisitions, specifications or data sheets should be laid out like books with main headings (chapters) for scope, standards, delivery and so on, in order to clarify where one topic starts and another ends. With the advent of computing technology most, if not all, of the documents concerned will be electronic.

THE MATERIAL REQUISITION

The material requisition is really a document that wraps up all the information required, to enable us to invite suppliers to tender for the goods or services that we require. Even organizations or individuals that do not have such a formality find themselves preparing a letter that performs the same function with the same information. The formality of the material requisition document helps provide information in a consistent and concise manner and avoids errors.

As already indicated, in a complex project there are likely to be three categories of requirements that will need to be requisitioned:

- Commodity materials – standard type materials that have a 'catalogue' type description.
- Items that are of a unique nature and require to be designed – designed items. These will require specifications and/or drawings to properly describe them.
- Services to cover the installation, commissioning and start-up of the unique designed pieces of equipment.

In general, the material requisition should be limited to one type of material or equipment and also to what will be supplied by one vendor.

The material requisition consists of a title page, revision record sheet, contents page and continuation pages. A cover page is addressed to the proposed tenderer and indicates the type of material required. The revision record sheet will indicate the issue number and the status of the requisition, as well as the date and initials for those who have prepared, checked and approved the document. The status will be indicated as; 'issued for quotation' or 'issued for purchase'. It will (as will all pages) indicate the page number of total pages together with the requisition number and revision number. A proprietary sample contents page is shown as Figure 7.1.

The continuation pages will be tabulated with columns for: item number; quantities or units; description; cost code; unit price and total price. In addition to the specific material requirements detailed under the description heading, other requirements will also be listed covering: drawings, documents, dimensions or other data concerning the equipment for checking by the designers; schedule information to enable expediting to be carried out; inspection and tests to be carried out; packing and transport requirements and spares. Attachments, listed under the description column, might comprise: specifications, data sheets and drawings.

Spares

Spares can comprise two or three categories: spare parts required for the operation and maintenance of a facility, and spares required as additional quantities as a contingency, say, during installation and commissioning. There may also be a requirement for specialist capital spares.

It may be acceptable when purchasing spares for an existing facility to purchase equal and equivalent components from a supplier other than the original manufacturer. However, this is definitely not allowed in the aircraft industry. The danger is when this is done without reference to the designers. On one project a shut down occurred due to the failure of a component in the main compressor. The manufacturers were summoned from Germany and presented with the evidence

				XXXX-1131X-XX	
				PAGE: 1 OF 14	
DRUM REQN.		**FOSTER WHEELER**		REV: **E1**	
EQUIP. ITEM NO		**MATERIAL REQUISITION**		DSN: XXXX	

Client :	XXXXXXXXXXX	Surveillance Grade :	1B
Project Title :	XXXXXXXXXXX	Document Category :	CLASS 1
FWEL Contract No : 1-XX-XXXX/X		Originating Discipline :	VESSELS
Project Location :	XXXXXXXXXXXXXX	Product Code :	1131X

REVISION	**E1**	E2	P1	P2	F1	F2
DATE	X.XX.XXXX					
ORIG.BY	A.A.A					
APP.BY						

CONTENTS

REVISION RECORD

1. SCOPE OF SUPPLY

2. TECHNICAL INFORMATION REQUIRED IN QUOTATION

3. ORDER OF PRECEDENCE OF DOCUMENTATION

4. QUALITY ASSURANCE & ENVIRONMENTAL REQUIREMENTS

5. DESIGN REQUIREMENTS

6. SCHEDULE OF SPECIFICATIONS, STANDARDS AND STANDARD DRAWINGS APPLICABLE TO THIS REQUISITION

7. GENERAL REQUIREMENTS FOR DOCUMENTATION.

8. INSPECTION/FABRICATION

9. PRODUCTION DATA BOOK – VESSELS

10. TRANSPORTATION/LIFTING

11. PREPARATION FOR SHIPMENT

ATTACHMENT ONE · QUOTATION DATA SHEETS (X SHEETS)

Figure 7.1   Material requisition contents

of the defective part, only to say it was not theirs. It was discovered that the stores manager had obtained the part from a cheaper supplier without reference to the maintenance engineers, causing a loss of several million pounds.

Documents, drawings and data

Documents, drawings and data can be a problem. Designers like to exercise their expertise and tend to ask for too many documents. However, preliminary drawings are essential to confirm the envelope size and termination points of what is being purchased. Other drawings will be required for incorporation in the operation and maintenance manuals.

Delivery requirements

Delivery requirements shown on the cover page of the requisition document should include a location for the delivery of the materials, or performance of the services. Delivery can be at the factory gate for you to collect or delivered to your door or a variety of stages in between. These are very well described and precisely described in 'The International Chamber of Commerce official rules for the interpretation of trade terms' – see Chapter 5.

SCOPE AND WORK DEFINITION

It is important to distinguish between what is to be provided – the scope of supply - the deliverables, and the extent of work that is to be performed – the scope of work. In mainstream project management the technique used to help define the product deliverables and the scope of work, and also overcome some of the problems of using words (see below), is known as the Product Breakdown Structure and Work Breakdown Structure.

The Association of Project Management defined 'work management' as:

'The process of breaking the project into manageable pieces of work. This can be achieved by first breaking the project into a product orientated family tree, that is a Product Breakdown Structure (PBS) and then breaking the project into a task orientated family tree, that is a Work Breakdown Structure (WBS).

The PBS is a product orientated hierarchical breakdown of the project into its constituent end items or deliverables without the work packaging or activities attached. It stops with the product end item definitions.

The WBS is the PBS with the principal work packages and activities needed to produce this. The WBS should depict a product in a manner in which technical accomplishment can be incrementally verified and measured, and provide the conceptual framework for all integrated planning and control of the work.'

A product breakdown structure is a formal and systematic way of defining the scope of a project. Of equal importance, the process helps to identify missing scope items and areas of ignorance. The graphical presentation is an end item subdivision of the project, product or item to be produced; it:

- displays and defines the product to be developed or produced;
- relates elements of work to each other and to the end product;
- enables responsibilities to be identified;
- forms a logical, structured and organized base from which to integrate the work to be done, the organization, and the planning and control system.

The product breakdown structure is developed by exploding the end product into its component parts and the services required (see Figure 7.2). The packages of work so formed must be clearly distinguishable from all other work packages. Each package is further subdivided into lower level elements representing units of work at a level where the work is to be performed. This process of breaking down the work is continued until the project is fully defined in terms of *WHAT* is to be done to complete the project. Although primarily orientated towards identifiable self-contained end products or deliverables, software, services and project management tasks may also be included. This work breakdown structure divides the work of the project into manageable units for which responsibility can be assigned.

Level 1:   contains only the quantified project objective or end item to be produced.

Level 2:   contains the major product segments or subsections of the end item. These segments can be defined by location or intended purpose.

Level 3:   contains definable components or sub-sets of the level 2 segments.

Level 4:   represents lists or units of work at the levels where the work is to be performed.

Another way of looking at the product and work breakdown structure is that the product breakdown structure deliverables are the nouns and the work breakdown structure level where the work is performed is the verbs.

Many projects can be defined in four or five levels. However, in aerospace it may take 11 or 12 levels to get to the level where work is to be performed.

If this technique can be used to represent the deliverables of a project together with the scope of work; it can also be used to illustrate the scope of the product or service that one wishes to purchase, together with the work that we wish the supplier to perform. To the scope must then be added the specification details defining the quality or features of the product.

SPECIFICATIONS AND STANDARDS

It is difficult to be precise about the difference between a specification and a standard, since the two terms are (to some extent) used interchangeably. A specification may ask for components to be in accordance with a particular standard. However, a standard may equally specify the composition

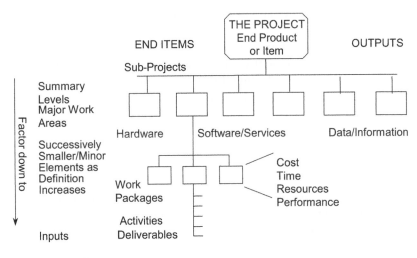

Figure 7.2   **PBS-WBS**

of the materials. Frequently used company specifications become company standards. Similar standards used by companies become industry standards (see below) and eventually they become national and international standards.

Standards are written compilations of best practice prepared and published by independent experts working under the sponsorship of national governments, for example, the British Standards Institute write and develop British Standards. There are also European Standards and International ISO Standards. Standards are also produced by recognized learned societies, for example, the American Society of Mechanical Engineers (ASME standards). When a standard is produced it deals with future developments not yet defined. A standard relates to an area of activity and is more general. Thus a standard, for a particular type of material, will itemise the constituent parts and identify the minimum strength required at various temperatures and so on. A specification, on the other hand, would say that this material must be used for certain aspects of the project and would reference the standard concerned. These standards do not, generally, have the force of law in their own right but may be brought into legal force through a related Act of Parliament, for example, The Clean Air Act or the Boiler Regulations. Insurance companies may also stipulate adherence to particular standards before granting the required insurance cover that may, in itself, be required by law.

Specifications are again a compilation of best practice but have no independent standing. When a specification is produced it relates to something that is known about or already exists. A specification relates to a specific object, a specific item of equipment, a particular project, facility or service. They are prepared by designers, engineers, architects, owners or consultants to represent the particular requirements for goods and services.

There are of two types of specification that are used together. First are the General Specifications, which expand on the standards to set out the requirements that an owner has identified and adopted. They are tailor-made to the originator's preferences but can be kept 'on the shelf' and reused (and kept updated) over periods of many years.

Second, are the Particular Specifications, usually referred to as Duty Specifications or Data Sheets. These stipulate in very precise format the exact requirements of the material, product, facility or service which is to be purchased. The tenderer must confirm compliance with the duty specification and provide an evident warranty of compliance.

Standards, though, have wide coverage and rarely can any one manufacturer or provider comply with all the stipulations contained in a standard. Similarly, a supplier may take exception to certain aspects of the general specifications. It is, therefore, important for the purchaser to identify the exceptions that will apply to standards and negotiate an agreement to the exceptions to the general specifications.

It is important to decide on the type or nature of the description of the goods or services that we require. Is it to be a unique document with all the details specified, or is it to be a functional or duty/performance type specification? Often it may be a mixture of both. In view of the number of documents involved it is also important to understand or define the order of precedence of the documents concerned as follows:

1. the particular duty specifications;
2. the general specifications;
3. the standards.

However, where Statute Law is involved the standards are likely to take priority.

All specifications are developed with a vision, in the mind of the designer, of the circumstances to which they apply. The more experienced the design organization, the more standard specifications will have been developed.

An example of a standard specification is shown in Figure 7.3[1] and illustrates the first two pages of the relevant specification. 'the purpose and use' described in its text is an excellent explanation of the objectives, basis and limitations of a standard specification.

However, standard specifications tend to be developed around the most severe situations contemplated. As a result all standard specifications are over-specifications – except for the worst

Process Industry Practices
Piping

PIP PN01CS1B02
Piping Material Specification 1CS1B02
Class 150, Carbon Steel, 0.063" C.A. Process

PURPOSE AND USE OF PROCESS INDUSTRY PRACTICES

        In an effort to minimize the cost of process industry facilities, this Practice has been prepared from the technical requirements in the existing standards of major industrial users, contractors, or standards organizations. By harmonizing these technical requirements into a single set of Practices, administrative, application, and engineering costs to both the purchaser and the manufacturer should be reduced. While this Practice is expected to incorporate the majority of requirements of most users, individual applications may involve requirements that will be appended to and take precedence over this Practice. Determinations concerning fitness for purpose and particular matters or application of the Practice to particular project or engineering situations should not be made solely on information contained in these materials. The use of trade names from time to time should not be viewed as an expression of preference but rather recognized as normal usage in the trade. Other brands having the same specifications are equally correct and may be substituted for those named. All Practices or guidelines are intended to be consistent with applicable laws and regulations including OSHA requirements. To the extent these Practices or guidelines should conflict with OSHA or other applicable laws or regulations, such laws or regulations must be followed. Consult an appropriate professional before applying or acting on any material contained in or suggested by the Practice.

This Practice is subject to revision at any time by the responsible Function Team and will be reviewed every 5 years. This Practice will be revised, reaffirmed, or withdrawn. Information on whether this Practice has been revised may be found at www.pip.org.

© Process Industry Practices (PIP), Construction Industry Institute, The University of Texas at Austin, 3925 West Braker Lane (R4500), Austin, Texas 78759. PIP member companies and subscribers may copy this Practice for their internal use. Changes, overlays, addenda, or modifications of any kind are not permitted within any PIP Practice without the express written authorization of PIP.

PIP will not consider requests for interpretations (inquiries) for this Practice.
Not printed with State funds

Figure 7.3   Standard specification

1   The cover page and explanation page have been reformatted on to one page for convenience. Reproduced with permission from PIP, Process Industry Practices.

circumstances for which they were envisaged. Specifications developed for a facility with a high fire risk may be unsuitable for a facility where the fire risk is small. A different example applying to petrochemical plants is the idea of 'double block and bleed isolation' – two valves in series with the space in between drained or vented through a further pipe with two valve isolations on it. This concept may be fine when applied to an 8″ line pipe, but less so to a 42″ line pipe where a valve is the size of a car. Consequently, the limits of the application of a specification need to be carefully explored and stated.

The young, enthusiastic or inexperienced designer may enhance a specification by choosing all the best features from the standard products of different makers without any reference to 'fitness for purpose'. This will most likely result in non-compliant quotations from most of the reputable suppliers, leaving a single acceptable quotation requiring special approval procedures.

On the other hand, designers have traditionally felt an urge to modify or add to a specification that they had used previously for a similar product. After all, if they do not do this they are not doing their job properly, or they are not demonstrating the importance of their specialization, and they are not keeping the specification up-to-date with the latest ideas. For example, in one facility the standard electrical specification for a compressor required a complicated delay mechanism in the vibration trip system – expensive both to buy and maintain. Investigation showed that this was the result of experience on another project where cost-cutting had gone too far. This had resulted in a rather flimsy structure that vibrated when the operating staff walked past changing shifts and tripped the compressor.

Designers tend to think that they know best what they require rather than trusting to the expertise of the vendor. There can be a desire by designers to tell equipment manufacturers how to design the components of the product they wish to buy. The result of this approach is over-specification and over expensive tenders. The trouble with over-specifying is that it inhibits the creativity of the supplier. They may be able to produce a more cost effective design. Work with the supplier to use standard components that are equally effective but cost less, and are faster to produce. Also, indicate where a supplier may suggest alternatives. It is also helpful, in order to obtain more competitive prices, to prioritize the requirements in a specification. Sometimes it seems that organizations have forgotten the definition of the technical role; namely, an engineer is someone who can do for £1 what any fool can do for £5!

With the advent of the CRINE initiative (Cost Reduction In a New Era) there has been recognition that if, say, a manufacturer has been in existence for a hundred years, they probably know their business and can be trusted to design components. There was a recognition that it was not necessary to define the details of the technology of a product you wanted to buy, but to define the objectives of what you wanted a product to do – a Performance Specification.

One of my earlier projects was a sulphuric acid concentration plant. Apart from its unique technology, it was a fascinating project due to the complexity of the material specifications. A major item was an extremely large vessel in which the weak acid was boiled. These harsh conditions required the vessel to be glass-lined and hung from the roof to cope with the expansion problems. For some reason, as project manager, I also had responsibility for specifying the vessel for purchasing. Realizing my lack of ability to carry out the design of the suspension ring from which to hang the vessel, I described how the vessel was to be used and the requirement for it to be suspended. Unfortunately, the glass cracked after the plant had been heated up. When the vendor was asked to replace the vessel their initial reaction was, 'It's not our responsibility'. I pointed out that the vendor had accepted performance responsibilities for their design and they replaced the vessel (at their considerable expense) without further argument. Sometimes it does not pay to try and be too clever and do someone else's job. In this case I left the design to the specialist.

One of the major weaknesses of people managing projects, and contractors in particular, is that they lack the experience of operating facilities. Another weakness of the contractor client, purchasing on behalf of an owner, is that they are only interested in the materials and equipment surviving for the contractor's guarantee period. Whereas, the owner client is more involved in the longer term perspective of operations and maintenance. Since suppliers learn from maintaining their equipment and guarantee its operation, they probably know more about the strengths and weaknesses in the design. Therefore, we should lean towards allowing the vendor to come up with solutions to our design problems. One of the key issues for success in projects is that, 'If you want a successful project you must involve the User in the earlier processes.' In this case the user of the specification or data sheet is the supplier. Consequently, we should invite the supplier to our offices to discuss how best to resolve the design issues, and present the specification.

All specifications for packages of equipment or complex items or systems should include words dealing with their reliability and availability of operation. Further, attention should be given to the interfaces with other equipment. In the early days of answerphones and facsimile machines it was necessary to have a 'fax switch'. The problem was, however, that the telephone worked, the fax switch worked, the answerphone worked and the facsimile worked, but all together the system never worked satisfactorily, and they all blamed each other!

Suppliers who are used to always providing items in a particular manner or combination will tend to assume, if nothing is said about them, that the items are included. It can, therefore, help to clarify the communication process by listing what is not included in the requirements. In effect, a negative specification.

Conversely, when dealing with a service the buyer will know considerably more about the location of where the service is to be provided. Consequently, if the buyer does not describe the features and services available at a particular location, and in particular the 'unknowns', then the tenderer is likely to make assumptions. These assumptions will tend to be of a pessimistic nature and, consequently, the tenderer will include additional monies, as contingencies in their tender, for the risks that they perceive – resulting in a more expensive tender. However, buyers can be reluctant to tell tenderers certain information due to its uncertainty – some of it may be wrong – and tenderers may then use it as a basis for making a claim for extra monies. It is in our interests to tell the tenderers as much information as possible if it removes contingencies from their tenders. Consequently, disclose the information, with whatever qualifications are necessary and state that it cannot be used as the basis for a contractual claim. Invite them to visit the location where the service is to be provided – see 'Tenderers Conference' section on page 96.

The main problem with specifications is that for the most part they use words to describe what is wanted. Documents that help describe one's requirements in a more structured manner and do not rely on words, to the same extent, are the data sheet and the material take-off. They can exist in their own right or form parts of the material requisition or specification.

The data sheet

The data sheet is the most straightforward form of technical communication and it helps to minimize mistakes. Most people working in a project environment are working under pressure and will, consequently, make mistakes or forget something. Organizations that regularly purchase similar types of goods develop standard forms – data sheets – to act as a checklist or memory jogger to ensure that items are not omitted. In addition, it ensures that the supplier marketplace becomes used to the format, thus enhancing the communication process.

An example of a less complex data sheet is shown in Figure 7.4. This example, and the particulars that follow, have been included to demonstrate the complexity of the detail required to define requirements for technology projects.

The title heading and the first three un-numbered lines of the data sheet form an identification header for each subsequent page.

Page two of this data sheet would have a similar format for the following information:

- fabrication details;
- inspection details;
- a nozzles schedule giving: size, rating, flange, stand out, elevation and so on.

Page three would be a full page for design notes and page 4 would be full page for a sketch of the vessel.

A data sheet for uncomplicated rotating machinery (say, a pump) would have a similar number of pages, but would be more complex and have sections for the following data:

- operating conditions;
- site data;
- materials required;
- performance required;
- driver type and details of driver;
- utility conditions;
- construction details with painting, packing and shipping details;
- heating and cooling requirements;
- bearings and lubrication details;
- instrumentation;
- weight and space requirements;
- spare parts and other requirements;
- piping and other connections;
- quality assurance, inspection and testing;
- design code references;
- welding and repairs codes and standards;
- material inspection requirements.

More complicated rotating machinery would have similar requirements, but in more detail, and would have additional pages for more complex requirements as follows:

- unusual conditions;
- noise specifications;
- vibration and axial position detectors;
- casing and other connections;
- allowable piping forces and moments;
- accessories comprising couplings, guards and mounting plates.

Materials lists – material take-off

A list of the material required can have a variety of names: Materials List and Bills of Material are perhaps the more generic terms. Bills of Quantity are used mostly in a civil engineering environment. As their names imply, these are lists of materials together with a description and the quantities required. There is little difference between buying stationery, or pipes and fittings for your central heating, or for a process plant.

VESSEL DATA SHEET			Attachment ONE		Page of	
Contract Number			Item Number			
Project			Service			
Location			Requisition			
Rev	REV E1	REV P1	REV P2	REV F1	REV F2	
Date						
Orig.						
Chck.						
App.						
1.	DESIGN DATA					
2.			Shell		Jacket / Coil	
3.	Contents					
4.	Design pressure (internal/external)			kg/cm²g		kg/cm²g
5.	Design temperature (max/min)			°C		°C
6.	Operating pressure (internal/external)			kg/cm²g		kg/cm²g
7.	Operating temperature (max/min)			°C		°C
8.	Hydrotest pressure/MAWP			kg/cm²g		kg/cm²g
9.	Minimum hydrotest temperature			°C		°C
10.	Contents S.G.					
11.	Corrosion Allowance (internal/external)			mm		mm
12.	MECHANICAL DATA					
13.	Layout		Loads			
14.	Vertical/horizontal		Basic wind speed		m/s	
15.	Vessel diameter	mm	Basic wind pressure		N/m²	
16.	Tan/tan	mm	Earthquake factor			
17.	Height of btm. tan line	mm	Base shear		kN	
18.	Type of heads		Base moment		kNm	

Figure 7.4   Data sheet

19.	Shell thickness		mm	AGITATOR		
20.	Head thickness (top/bottom)		mm	Download	kN	
21.	Type of supports			Torque	kNm	
22.	Insulation thickness		mm	Bending moment	kNm	
23.	Vessel volume		m³	WEIGHT		
24.	Liquid volume		m³	Shipping	kg	
25.	Jacket/coil volume		m³	Empty	kg	
26.	Heat transfer area		m²	Operating	kg	
27.				Hydrotest	kg	
28.				Internals	kg	
29.	MATERIAL DATA					
30.	Shell			Bolts (external)		
31.	Heads			Nuts (external)		
32.	Jacket/coil shell			Bolts (internal)		
33.	Jacket/coil head			Nuts (internal)		
34.	Nozzle flanges			Gaskets (external)		
35.	Nozzle necks			Gaskets (internal)		
36.	Nozzle pads			Internal baffles		
37.	Girth flanges			Internal pipes		
38.	Supports (top at shell)			Internal flanges		
39.	Supports (lower)					
40.	Insulation supports					
41.	Stiffening rings					
42.	Platform clips					
43.	Tailing Lugs					
44.	Lifting Lugs					

FOSTER WHEELER ENERGY LIMITED

Changes are denoted by a revision triangle Δ in the right hand column.

Figure 7.4   Concluded

The items are identified with catalogue type descriptions and any appropriate standard specification numbers. As stated in the standard specification Figure 7.3, it should be stated that brand names are for illustrative purposes and that they can be substituted with equivalent types from other manufacturers. It is also sensible to encourage tenderers to submit alternative proposals that they consider would be more economic, provide improvements or benefits.

Whilst the descriptions may be relatively straightforward, the quantities are one of the most difficult things to get correct. It is, therefore, prudent to allow for some surplus for commonly used items. Allowances have to be made for wastage, loss and installation and commissioning spares. This results in over ordering, asking the supplier to accept the return of surplus quantities, and eventually, to one's embarrassment, requisitioning a final few more!

It follows that it is important to make sure that you include a clause allowing you to return surplus material to the supplier for the same price; adjusted naturally for the cost of transport. Figure 7.5 illustrates a manual material take-off sheet and again demonstrates the extent of the detail required to define a project's requirements. This was the 'Kellogg Way' prior to its incorporation into a computerized Integrated Project Management System – a full Engineering, Procurement and Construction Material Management System with all the KELLWAY features and more.

Administering the documents

In order to facilitate any discussions, for whatever reason, all of the foregoing documents need to have identifying names and unique reference numbers, revision letters, dates and page numbers. These should appear as a header on each page of the document concerned. Your company's name, logo and the identifying form reference can be placed as a header or footer. The cover page should, of course, have the name of the company to whom the enquiry is being sent.

WORDS, WORDS, WORDS

Everything that has been said so far can apply equally to the supply of goods or services. However, when it comes to the provision of services we are far more likely to be starting with a blank sheet of paper. In this situation we are dependent on our ability to describe what we want, using words. The problem of defining what one wants is best illustrated as follows:[2]

Why is my kitchen floor not level? Because I did not ask for it to be dead level. I anticipated some problems in renovating my house, but I never expected to have a metaphysical discussion with the builder on the essence of levelness. Everyone agrees that the floor has bumps but the builder swears that it is level. It is just not dead level. Semantics are as important to the modern builder as any knowledge of construction.

Before starting I spent a long time preparing what I thought was a detailed specification. The builder priced it and, after some haggling, we agreed a contract sum. I now realize that the reason he took so long was that he was scrutinizing the specification for examples of vagueness and omission. These were expensively priced as 'extras' and were presented at the end of the project.

The window was the first indication of problems. The specification required the removal and disposal of a window and the installation of a new one. I thought I had been astute to include the removal in the specification. You and I might think it usual to find glass in a window. Surely, the whole point of a window is that it should have a pane; without the glass the hole might as well be bricked up. A builder never expects a window to have a pane in it, unless it is defined in black and white. Glazing

---

2   Adapted from an article by John Crace in *The Independent on Sunday*, 23 June, 1991.

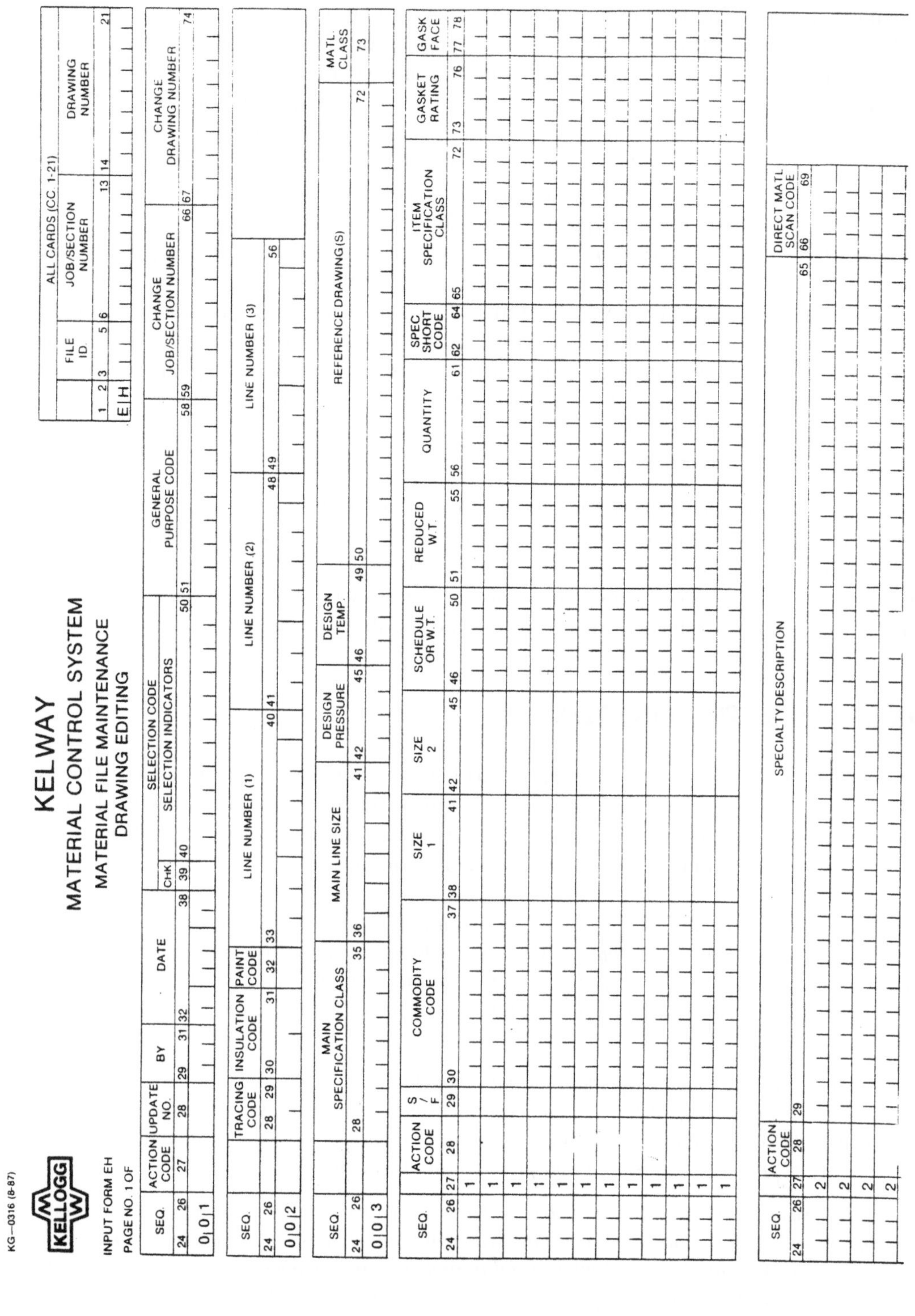

Figure 7.5   Material take-off

is extra. The doors were another problem. We had agreed a price of £100 per door. This had seemed very expensive but I was assured that the doors would cost about £70 and the labour would be £30. Some doors may cost £70 but not ones that are made of hardboard stapled to a softwood frame. I discovered that these doors cost £15. Now we were in very murky waters. We had agreed a contract, but is £15 about £70? What does 'about' mean? By the way, never expect a door to have handles unless you ask for them.

The building works have now stopped. I will not pay the builder until he finishes the works, and he will not do any more work because he thinks he has finished. Naturally, we cannot agree on the meaning of 'finished'.

In the meantime, my understanding of words has improved, and my kitchen floor is still not flat.

I have experienced this myself. Relatively recently I undertook two major projects – building a conservatory and rebuilding a garage. Whilst the tiles were being laid in the utility area of the garage I observed that the tiler was only using a two metre piece of timber to level the floor. Being concerned about the floor being level, I checked it the first evening with my long spirit level – to my surprise it was dead level. The following morning I asked the tiler, 'How did you get the floor level with only a long piece of wood?', 'Oh, it's not level, it's flat' was the response. 'In that case I would like the conservatory floor 'flat and level,' I stated. 'That will be more difficult' was the reply!

This illustrates that the difficult part in defining what one wants is that one has to use words. As Churchill said, 'Tricky little buggers.' Unfortunately, as illustrated above, words mean different things to different people. However, experts have a precise, or know the precise, meaning of words in their own particular field. People will interpret the words within their frame of reference. A buyer will use words within the culture of their business context, but the supplier has a different culture and business context and the words can mean something different. In technical environments it should be slightly easier because technical words have been developed that have more precise meanings, and in many cases one is dealing with numerical descriptions of weight, height, length and volume and so on. However, with more complex items the technical data will need to be complemented with descriptions and, unfortunately, this is where the technologists' skills are weakest.

Another problem for the less experienced person is the tendency to assume that certain features, that are always associated with a particular item, are included. Doors have handles! For example, my sister recently purchased a bath that had been annotated, in a whole list of items, '*vidanges pas inclus*'. Not being familiar with technical words in a foreign language, when purchasing in a country other than one's home country, can be a minefield. Not surprisingly, the bath arrived without any drainage connections. Conversely, if one is purchasing internationally one needs to be aware that the suppliers may not be as familiar with the English language.

Writing and communicating clearly what is required can often be seen as a chore. As a consequence, the process may not be done with the care and attention that it requires and sloppy wording may creep in. In addition, because we are so familiar with our own project and areas of expertise, unintentional omissions occur and we take items for granted. Further, as indicated above, technologists who have the dominant input to specifications are not known for their ability to write English effectively. We may make statements that have more than one meaning. In this case the tenderer is likely to be confused and may make an incorrect interpretation. In our desire to be careful we may state requirements more than once. If the words are not identical we have merely added more words and caused confusion. Repetition to look out for is where information is in the body of the text of the specification as well as part of the numerical data on the data sheet.

The more knowledgeable person will describe every feature in detail, in a desire to make sure that everything has been covered. Unfortunately, this leads to the problem that 'the more you write the more you leave out'. Having comprehensively listed all the features that you desire, if you do not detail something, then clearly, you did not intend it to be included.

The more the details of how a product is designed are covered by national or international standards, the better. If the appropriate ones are used it prevents us from over-specifying. The advantage of standards is that the words have already been checked. They have been put together in a way that is acceptable to all parties and have acquired a common meaning over time.

A case[3] drawn to my attention by David Wright illustrates the specification writer's problem. Summarized simply the scope of work stated:

The contractor will perform:

```
┌─────────────────────────────────┐
│                                 │
│       Work A Description        │
│                                 │
└─────────────────────────────────┘
```

and

```
┌─────────────────────────────────┐
│                                 │
│       Work B Description        │
│                                 │
└─────────────────────────────────┘
```

in accordance with the specification.

The question is, what does this mean? Work B was performed satisfactorily in accordance with the specification. However, it was extremely difficult to perform Work A. The contractor did their best but it did not conform to the specification. The judgement given in the Technical and Commercial Division of the High Court was that Work A was not required to be performed according to the specification. Using a strict grammatical interpretation of the words the judge ruled that the specification said: (perform Work A) and (perform Work B in accordance with the specification). If A and B had to be in accordance with the specification then the words should have said '*both* A and B...'. I understand that each letter of the word 'both' cost £1m!

Being aware of these problems, we check our own work but do not see the obvious mistakes. Because we are so familiar with the material we see what we know should be there rather than what is actually there. People other than the writer should check the documents. People that Meredith Belbin calls 'Completer Finishers'[4] are needed. Belbin describes completer finishers as typically, 'Painstaking, orderly, conscientious.' They have the following positive qualities, 'A capacity for follow-through. Perfectionism.' He describes their allowable weaknesses as, 'A tendency to worry about small things. A reluctance to let go.' All of which look like positive qualities to me in this context. What you need are people who are used to checking software, if you have them in your organization.

QUESTIONS AND THE TENDERERS CONFERENCE

Unfortunately we will have done a less than perfect job of communicating our requirements, and in any case people will interpret words in different ways and will make assumptions. Consequently, there will be questions during the enquiry period. Is something included or excluded? It is then important to respond in such a manner that it is seen to be fair to all tenderers. It can be useful to ask tenderers to submit questions say 1 or 2 weeks after receipt of the enquiry documents. The

---

3   Tarmac Roadstone Ltd *v's* MathewHall Ordtech Ltd. In the second half of the 1990s.
4   One of the eight team roles identified by Meredith Belbin's Self Perception Inventory.

questions, together with the answers, can then be circulated to all the tenderers, so that nobody can identify who asked which question, but they all have the same information.

Questions relating to the provision of material and equipment can be dealt with by written answers. Equipment supplies may require the supplier to visit the location where the equipment is to be delivered. In this instance invite all the suppliers, and if necessary their transport subcontractors, on the same day, explain any restrictions and answer their questions in open forum. You want everyone to get the same message so as to keep their costs, and thus their prices, to a minimum.

For the more complex service requirements, the tenderers should be invited to a clarification meeting at the location where the service is to be provided. Should we hold this meeting with all of the tenderers present at the same time, or should we schedule separate meetings for each company in turn? Obviously it is more efficient and involves less effort if they are all invited together. However, will they really ask the question that is worrying them? They may not want to show their lack of sophistication in front of the competitors and, as a result, provide a less than satisfactory tender.

Alternatively, some tenderers may play games by asking questions that they already understand in order to put off other tenderers. Even if we ask tenderers to send in their questions beforehand and prepare the answers, the more astute tenderers will be watching peoples' body language in order to identify who asked which question.

The professional approach is to request the questions early enough to be able to hand out the written answers at a meeting. Schedule separate meetings with each company and, as you show them your premises and facilities, develop a relationship with them. Are these people you could work with, and are they keen to provide the service you require?

8 *Selecting the Tenderers*

This topic is part of the process of choosing the right supplier. It is not complex or difficult but it is one of the most important. If the wrong name(s) is on the tender list then problems are bound to occur later. In the same way that the successful execution of a project depends on the choice of individuals to form the core project team, so it depends on all of the other participants. The contractors, consultants, equipment manufacturers and material suppliers are all important when choosing the larger outer project team.

The subject can be divided into two parts. Firstly, deciding how to select those for the tender list and who to put on it. Secondly, pre-qualifying those on the tender list in order to decide whom to invite to submit a tender. The key to the whole process is experience.

I was fortunate on joining a company to be appointed as the proposal manager for an enquiry, and then to be the project manager when we were awarded the contract. We had worked out what and how most things were going to be organized during the tendering phase. Upon award it was important to maximize the advantage of having the same project team as the tender team and implement all of the administrative issues as rapidly as possible. Consequently, I was busy signing off lots of documentation, but then the project procurement manager gave me a whole stack of procedures. I was reluctant to rubber stamp the documents without being aware of their contents. In any case I was new to the organization and the procedures would give me a good insight into how the company worked. Consequently, I took the most unusual step (for me!) and took them home to read overnight. Whilst I was satisfied with the content of the vast majority of the procedures there were two or three issues where I thought; if I sign off on this, I will not know what is going on and, I will lose control. The primary issue, that I still remember, was that as project manager I wanted to authorize the tender lists before enquiries were issued. I wanted to ask why certain people were on the list, who had suggested the name and why, what experience did we have of the companies' performance or other relevant issues in controlling the budget and schedule and meeting our objectives?

THE TENDER LIST

Effective procurement is crucial to completing projects on time and within budget. It is about knowing who the right supplier is, who will provide the right material of the right quality, and knowing who can be relied upon to submit acceptable offers and perform effectively (the right price with the right delivery).

However, the proposed tender list cannot be formulated until a key strategic decision has been made. Is the enquiry to be sent to local, national or international companies? Local companies may be preferred when, say, a short call-out time is required. An international enquiry may be preferred when the client wants to take advantage of the tenderers' ability to raise local finance.

Finance is the other strategic decision that will affect the tender list, particularly for large contracts. Is financing to be part of the competitive process in order to take advantage of different national loan arrangements? If credit financing is to be part of an enquiry then it may not be necessary to invite tenders from more than one supplier per country, since the loan arrangements will be similar. If suppliers' proprietary processes are involved, or there are other technical reasons, then the financing arrangements may make a difference to the tenders.

If the project as a whole is financed by international loans, the nature of the loans concerned may dictate what materials are to be sourced in which country. I have never forgotten when, on one of my projects that had three international loans, we bought piping insulation material from India using our dollar loan whilst we had an excess of rupees on our Indian loan.

Potential tender lists are generated from a number of sources. The functional department responsible for the technology involved, the project procurement department (who will access the company procurement department database), project engineers and even, sometimes, the project manager. In the unusual situation when there is a complete absence of experience or any other information or data, one is forced back to basics, and today the Internet will be the first place to search. However, do not forget to peruse the 'yellow pages' of the particular business sector or industry; specifically, the professional institution journals and other trade publications. Further, do not forget company brochures and annual reports.

In whatever manner the list of names has been compiled, a company's website will also be the primary source of data. Comparable information and data should be collected on the technical, commercial and financial capability and suitability of the suppliers or contractors. The companies on the list can then be examined on a like for like basis. Typical information might be:

- ownership of the company;
- financial stability;
- company size – turnover;
- geographic location;
- extent of product and/or services offered;
- projects performed;
- technical expertise;
- manufacturing capacity or resources available;
- membership of trade and industry associations;
- contact details.

Despite much information being readily available on the respective organizations' websites there may still be unanswered questions. The recent history of corporate failures and high profile cases such as Barings and Enron have highlighted the need for risk analysis and assessment of the financial health of all companies, even the larger organizations. Consequently, it may be necessary to issue a Request For Information (also known as an RFI). A request for information will be simply a list of whatever questions that you, the client, wish to have answered, in order to fill in the blank spaces of the information available.

In general, any company that has been in business for a while, either awarding or performing projects, will have a purchasing capability. The purchasing or procurement department should have collected information about the performance of companies that they have used, and should be researching the market place for new suppliers or contractors.

If the sales and marketing personnel of suppliers and contractors have done their jobs properly, you should already be conscious of companies with relevant capabilities. They need to be aware of you as much as you need to know about them.

As a generalization, contractor clients (working for more and different owners in different fields) tend to have a broader information base than owner clients. However, owners have the vital operating and maintenance data.

Back in 1981 Bechtel already had a well established Supplier Information System, 'A computerized data base of 6000 suppliers and contractors throughout the world who offer some 2500 commodities and services.'[1]

It is interesting to note that it was a further 15 years before a leading owner organization created a similar database.

[2]*It was called the 'billion dollar challenge'. In 1996 BP, the international oil group, decided that it could save $1 billion from the $15 billion a year that it spent on goods and services.*

The key was information. BP could only make the savings by getting an overview of all its spending decisions. Once it had the full picture, it could negotiate better deals with its main suppliers.

But two years ago, identifying those opportunities was difficult as there was no single place in which all the purchasing data was gathered.

There was a need to build a computer system to collate information about who was buying what from whom. BP tackled this problem by building a data warehouse - information from many sources in analyzable form.

Users as far afield as Australia, Alaska and Azerbaijan may interrogate the system to find if a potential supplier is doing business elsewhere in the group or if another supplier offers better terms. BP procurement managers may use this information to aggregate purchases and to negotiate better terms.

An important consequence is that BP can rationalize its supplier base. 'It could easily be seen as a big stick to beat suppliers,' But...it allows suppliers to develop a more collaborative mode of working with BP.

'The real value is in the second tier suppliers, '...'For heads and procurement, it is pretty obvious who are the top 20 suppliers. But look one level down, they don't actually know who are the next 20.'

Data from 20 countries is in the system, now used by 700 people all around the group.

Within a year of the system's introduction, savings of at least $15 million were made; equivalent to five times the project's cost.

Regular field research during quiet periods of work will save time when projects are active. Field research is essential if the organizations on a proposed tender list are unknown. Visit the companies, tour their facilities and meet key personnel. Meeting the managing director and senior managers helps to evaluate the management culture and provides personal contacts for future use. Carry out a Quality Assurance system audit. Ask for references, go and see some work that they have carried out and talk to their recent clients (projects, operations and maintenance personnel).

1   *Shopping the Worldwide Market*. Bechtel Briefs July/August, 1981. Published for the employees and friends of the Bechtel group of companies, San Francisco.
2   'BP overcomes fear of the "too difficult" box', *Financial Times*, April 15, 1998.

Much of this work can be simplified in the process industries by subscribing to an organization called 'First Point Assessment'.[3]

First Point Assessment Ltd (FPAL) is a non-profit making oil and gas industry owned and governed company established in December 1996. Its purpose is:

For purchasers: to provide objective information on the potential and actual capability of suppliers and contractors.

For suppliers: to enable suppliers to provide consistent and up-to-date information to potential purchasers in a cost effective manner.

All parties to minimize inefficiencies between major purchasers and suppliers through eliminating duplicated registration and assessment activity.

To provide opportunities for improvement throughout the supply chain through enhanced knowledge of strengths and weaknesses.

First Point Assessment essentially collects information on contractors and suppliers (Registered Suppliers), recording their capability over a number of different attributes in a central database (FPAL). The data is expressed graphically as company 'Profiles' and is collected by coordinating feedback of actual performance (Performance Feedback) and by conducting an assessment from the completion of a separate capability questionnaire. The database is made available to potential purchasers (Subscribers) who are subject to appropriate restrictions. Benchmark performances are made available to individual Suppliers and Subscribers.

The Freedom of Information (FoI) Act can also be used as a mechanism to gather information. *The Financial Times* reported on August 30th, 2005:

There are signs that the private sector is increasingly using the FoI regime to extract information about bids and contracts from public bodies such as local authorities and NHS trusts.

But many companies are reluctant to tell the authorities that they are seeking the information for fear of damaging their relationships with officials.

...companies are using private e-mail addresses or employing consultants to make requests for them.

Whilst, at first glance, this may appear to be of main benefit to a supplier or contractor (which it is), it is also useful to see who other clients chose for their projects and how the successful tenderer performed.

With all of the data available from the corporate list of suppliers and contractors, it is now necessary to make choices for a project tender list.

The project tender list

The project list needs to be compiled as early as possible. The objective of the project tender list is to determine the best choice of suppliers and contractors for the subject project.

It would be a mistake to compile the initial project tender list, covering all the goods and services required, in a routine manner. Your past experience may not be as relevant for an innovative project. It would probably be a mistake to use a high technology contractor for a low technology project – they will over-design and be too expensive. Further, a project manager might be reluctant to choose

---

3   See *Additional Sources and Contact Details.*

new or unknown suppliers for a critical high risk project. New and unknown companies should be given opportunities to prove their capabilities on smaller, low risk repeat type projects.

Naturally, the list will depend upon the goods and services required. A supplier of standard products may not be appropriate if the project requirements are tailor-made to very specific requirements. Similarly, it may not be sensible to include someone whose capability is only peripheral to the technology of the project. Contractors like being able to extend their capability at the expense of their clients. There are times when this is acceptable but other times when it may put the delivery of the project at risk.

Suppliers of materials and equipment that are on the project's critical path need to be chosen for their reliability and proven track record of delivering on time. Materials and equipment that has significant float (spare time) within the project schedule can be chosen differently. For example: a manufacturer may produce good quality goods but have a poor delivery record.

At the stage of preparing the project tender list it is useful to survey the market place to find out what other projects are underway, and which suppliers and contractors are busy with other work.

The design department should have indicated the types and approximate quantities of materials and equipment that they anticipate being required. Similarly, the execution departments (manufacturing, installation or construction) should indicate the type of services needed. If not, then purchasing should take the initiative so that a start can be made on developing the project tender list.

If a contractor is working on behalf of an owner client then the owner may well wish to include or exclude certain names from a tender list. For example: in order to standardize on certain types of equipment. Rationalization of spares may dictate that a particular vendor's equipment should be chosen. Further, as a result of their own experiences, the contractor may also wish to exclude specific names from the tender list. In these circumstances it is unlikely that an owner client will override a contractor's recommendation. In addition, the owner client may well want to approve the proposed tender list for each enquiry. In order to reduce the bureaucracy the contractor may be given discretion to operate independently for purchases below an agreed value. It is sensible for the owner client to perform occasional checks to ensure that the contractor is not misusing this delegated authority.

Purchasing will then contact suitable suppliers who have experience in the particular type of materials, equipment or other work, to check whether they have the capability to submit a good competitive price and the capacity to achieve an acceptable delivery, for the proposed project. This process is called pre-qualification.

Number of tenderers

The short list of tenderers should consist of between three and six potential tenderers. The number should reflect the fact that a tenderer may pull out at the last moment, or that a tender may be submitted that is unrealistic. To some extent this is a failure of the pre-qualification process. Nevertheless, it can still occur.

Tendering is an expensive process and only one tenderer will be successful. If there are too many tenderers they may feel that they do not have a realistic chance of being awarded the contract and may not submit realistic prices. If too few responses are received then the client faces the cost and, more importantly, the time involved in repeating the process.

Public bodies and international institutions will have specific requirements about the number of tenderers. However, the minimum number must be set to ensure that there will be at least two viable tenders. As already indicated, this is in case a tenderer withdraws, or in case it is not possible to reach agreement with the selected 'winning' tenderer.

There are a number of options available when you are unable to identify a sufficient number of potential tenderers:

- advertise;
- go international;
- ask the design department to break the work down into smaller elements;
- talk to the design department about redesigning the requirements;
- in conjunction with design, develop a potential supplier to become competent.

International issues

When purchasing internationally it may be necessary to place an advertisement in some key publications. Project funding needs may make this a requirement or there may be insufficient local suppliers or contractors. The advertisements will need to be in the local language and will provide brief details of the proposed project and the materials and or services required. It will invite relevant suppliers and contractors to apply for a pre-qualification questionnaire. The pre-qualification process often corresponds to, what is entitled, requests for Expressions of Interest. Since any interested contractor or supplier is able to apply, it gives new or unknown tenderers an opportunity to compete for the work.

The following is a representative selection of the issues covered in advertisements in the financial press in 2006/7. The wording, though, has been modified for simplicity. However, the combination of issues and peculiarities of clients' specific requirements are extremely diverse. Any reader who wishes to study the details of the international tendering process further is recommended to peruse past and future advertisements over a significant number of months.

- [The Company] is inviting interested parties to submit an expression of interest to obtain the Request For Qualification information (RFQ). [The Company] reserves the right to reject or accept any expression of interest received on the basis of the criteria set out in the RFQ.
- Expressions of Interest (EOI) for pre-qualification for the supply and installation of equipment. Tenderers must:
  - be experienced in the type of project;
  - be experienced in the country region;
  - have people experienced in the type of project and in the country region;
  - sound financial references;
  - client and consultant references for projects completed;
  - demonstrate that the company has worked as a subcontractor for a main contractor.
- The tender documents shall be issued to tenderers who submit definite proof of owning plant [used to produce the material required].
- [The Company] invites expressions of interest from contractors wishing to qualify to tender for engineering, procurement, construction, testing and commissioning of the following works... These will be executed on cost plus design and build contracts.
- Pre-qualification is open to suitably qualified and experienced international contractors. Contractors who fulfil the following requirements shall be eligible to apply:
  1. A minimum average turnover over the last three years of...
  2. Experience in *xxx* project type in the last 5 years.
  3. Minimum net worth of $ *x* m.
  4. Successful track record of timely completion of projects.

- In the case of foreign consultants, they will have to associate themselves by making a JV/consortium with [national] firm/firms having experience in any of the areas.
- Interested contracting companies are requested to submit their Expression Of Interest (EOI) with a) company profile along with contact details/communication address; b) details of infrastructure and equipment; c) references of similar work done in the recent past and; d) financial details and other information that the company feels is relevant and may qualify them for being considered by [the company]. These details should be sent within 10 days from the date of this advertisement through e-mail and supported by a hard copy through courier.
- [The company] is interested in lease-in of three (03) [XXX] aircraft fitted with [XYZ] Engines on dry lease basis for a period of 5 years extendable to 7 years, from established leasing companies/owners/operators. Offers from agents/brokers will not be entertained.
- [The company] invites EOI. Detailed EOI document and required formats for data submission can be downloaded from the Tender Notice Section of [the company] website.

The embassies or consulates in your home country (for the prospective supplier country) should also be contacted since they should be able to suggest a suitable list of tenderers for the proposed commodity or service.

United Nations projects or public sector projects in the European Union will all have their own set of rigorous rules, that will need to be followed, for how and who to invite.

As already indicated, projects using loans from banks are similar. The bank will dictate where certain materials should be bought. After all, they make the loan so that the money should be spent in a particular economy. The International Bank for Reconstruction and Development (IBRD) and International Development Association (IDA); The World Bank states:[4]

...four considerations generally guide the Bank's requirements:

    (a)   the need for economy and efficiency in the implementation of the project, including the procurement of the goods and works involved;

    (b)   the Bank's interest, as a cooperative institution, in giving all eligible bidders from developed and developing countries an opportunity to compete in providing goods and works financed by the Bank;

    (c)   the Bank's interest, as a development institution, in encouraging the development of domestic contracting and manufacturing industries in the borrowing country; and

    (d)   the importance of transparency in the procurement process.

1.3   The Bank has found that, in most cases, these needs and interests can best be realized through International Competitive Bidding (ICB), properly administered, and with suitable allowance for preferences for domestically manufactured goods and, where appropriate, for domestic Contractors for works under prescribed conditions.'

The following exercise illustrates the difficulty of choosing the right organization for the proposed tender list. So as to avoid qualifying your choice by stating that 'it depends upon the price' assume that (for the purposes of the exercise) there is nothing between the proposed suppliers in terms of price or commercial terms. However, in reality at this stage (preparing a tender list) you would not be aware of this commercial information.

---

4   *Guidelines*, procurement under IRBD Loans and IDA Credits The World Bank, Washington D.C. USA Page 2. Revised January, 1999.

WHICH WOULD YOU PICK? by David Wright

You propose to place an order for the first part of what is likely to be a substantial purchasing exercise for new quality control and sampling and testing equipment. The project is high profile and the decision has been made to go for 'state-of-the-art' equipment. The technology is still slightly experimental, but the design technologists are insisting, and something new is needed anyway.

You are considering three companies for the contract, all of whom appear to meet your specification.

Company A is a prestigious manufacturer based in Tokyo, but with a small sales and service company based in Reading. They have a reputation as a totally reliable, but very conservative organization that does things well but in their own way. They are the main supplier to the Japanese market and have a number of active contracts for similar equipment worldwide.

Company B is a large UK manufacturer which has done roughly similar work in the past. However, the company has been doing rather badly over the last few years and has had a bad press recently. They have said that they are very keen to develop a long-term business relationship with you, and their technical competence is not in doubt. However, they are known to adopt a very aggressive marketing and commercial approach to customers.

Company C is a small, rapidly growing local company. Some of their technical staff are personally known to you, and some have actually been employed by you or by Company B in the past. The company has gone for growth and new technology and picked up some good contracts. It appears to be doing very well.

Which is your preferred candidate for the work and what are the factors that influence your decision?

Discussion

This is all about perception and decision making. Theoretically decision making involves collecting pertinent and timely information and using your experience and judgement to arrive at a conclusion. Thus when we first look at a new supplier we make a decision about what type of company they are (in theory), using all the available marketing evidence. In practice, this is then modified by our instinctive feelings (experience and judgement).

Whilst suppliers A, B and C do not really exist, they represent the most common results from market research into what purchasers look for in a successful supplier.

A is the company in control of the technology, high prestige, government contracts, big research facilities, selling up-to-the-minute technology. The comfort factor that the buyer gets is the knowledge that the equipment will work.

B is the large group of companies. Their size and bargaining power with subcontractors and their staying power provides two comfort factors: price control and stability. They will be there when the buyer needs spares and servicing.

C is a small helpful user-friendly company. They are ready to leap into the future and the buyer remains in control.

The disadvantages to the three types of company are:

A is Arrogant – they will give the buyer what they know the buyer ought to want, rather than what the buyer thinks they need. Company A is also conservative in that they will offer the buyer a proven solution.

B is a Bastard – commercially and contractually adept. They are prepared to drive a hard bargain and perhaps to bully or blackmail their way out of trouble.

C is a Child – vulnerable to disasters. They have fragile finances and they are short on resources if things go wrong.

In asking training course delegates to pick A, B or C the usual mix is about 30 per cent A, 10 per cent B, and 60 per cent C.

David Wright's advice is that psychologically we start to deal from our preconceived view of a company. We then get to know people and establish a working relationship. However, if the contract goes wrong, so that we are under psychological pressure, then the way that we manage the relationship or contract will inevitably be coloured by our preconceived ideas.

PRE-QUALIFICATION

The purpose of a pre-qualification questionnaire is to request information in sufficient detail so as to be able to select the companies to be invited to submit a tender. The questions must be sufficiently probing in order to eliminate unsuitable tenderers. Each pre-qualification questionnaire will have some common generic questions and some very specific questions focused on the project in question.

If a request for information has already been used to compile the list of potential tenderers for the project, then there will be fewer generic type questions and more project focused questions. However, if a request for information has not been used then the pre-qualification questionnaire may also serve to gather information and data.

Pre-qualification questionnaires can vary from one or two pages to 15 pages or more. Sometimes it appears, to the supplier or contractor, to be a process of free consultancy! The questions must be answered. Since, if the proposed tenderers do not pass this test then they do not receive the enquiry documents. However, the process should be matched to the scale and complexity of the project. It is unreasonable to put suppliers through expensive and complex qualification processes, on each and every occasion, for routine projects.

Thus, for standard repeat type supplies of materials and equipment, and in order to reduce the amount of work imposed on suppliers, it is not uncommon for clients to carry out single annual pre-qualifications.

The pre-qualification document will consist of facts and information about the project and a series of probing questions. The following information should be provided:

- the client's name and contact details;
- the project title and location;
- an indication of the size, capacity or dimensions of the project;
- any unusual features;
- the proposed contracting strategy;
- a brief description of the proposed work required by the supplier;
- the intended programme: contract award date, start of work date and contract completion date;
- the expected enquiry issue date;
- the intended date for the submission of tenders.

There are two principal areas of questioning. The first is simple and straightforward. Is the supplier or contractor willing to submit a tender on the terms indicated? There is no point issuing an enquiry for fixed prices if market conditions are such that suppliers are unwilling to tender on this basis.

A variant of this question can sometimes be asked at this stage or at the earlier request for information stage. I remember receiving a visit from a major client to discuss their proposed project and during discussions they asked, 'What type of contract do you think we should use?' I was well aware that they had their own very clear ideas. Consequently, this was a test question. At the time the company I worked for was very risk averse and I am sure the company response would have been, 'We prefer reimbursable contracts.' However, I decided to give a personal response and said, 'As a project manager I much prefer fixed price contracts.' This was the right answer. Why should a client let a contractor play with the client's money if the contractor is not prepared to risk their own money? Will they look after your money as if it was their own?

The second area of questions needs to be more probing. Does the proposed tenderer have the experience, expertise, capability, resources and financial stability? These are the factors that *qualify* an organization to be on the tender list.

As already indicated above, the pre-qualification questionnaire is similar to an exam. If the proposed tenderer does not answer satisfactorily then they will not pre-qualify and will not be issued with the enquiry documents. However, it is surprising how often people do not respond to the questions that have been asked but answer questions that they would like to have been asked! This is quite irritating since time is wasted chasing up the right information before the analysis can start.

How people respond to a pre-qualification questionnaire reveals a lot about the sophistication of their organization, understanding of the issues and potential behaviour on the project.

Typical information, together with explanations, that might be requested for suppliers of services is as follows:

1. Please confirm that your company will submit a tender in accordance with the proposed contracting strategy and in accordance with the attached terms.

It is important that the contracting strategy is very clear, since this is a make or break request. If the supplier or contractor cannot reply in the affirmative then they do not pre-qualify. The remaining responses then become irrelevant. However, some contractors try to have their cake and eat it by keeping their options open with phrases such as, 'We are willing to tender on any terms that are appropriate to the project.' If this is a response to a request to confirm their willingness to submit a tender on a fixed price basis, then it reveals an element of risk averseness. They have failed to realize that this is not an Invitation To Treat, and so their responses are not an offer. At this stage there is no contractual commitment in responding with a positive attitude.

2. Describe your capabilities, experience and resources with XYZ technology.

Every project has some facet of technology that is crucial to the project. It may or may not be very complex but it will be essential that the supplier has the necessary capability. Consequently, this can be another make or break situation if the tenderer fails to be convincing in their response.

The technology may involve a specialization where the tenderer has only one person capable of doing the work. In these circumstances you should question what happens if the specialist becomes unavailable through sickness or the demands of other projects.

3. List relevant and verifiable current and prior experience with similar projects undertaken in the last 3 (or 5) years. Provide their cost and schedule performance figures. Give reasons for any disparity between the original contract figures and final costs and schedule achieved.

This response should provide factual data that can be verified. This request for an experience list is limited to 3 years due to the turnover in personnel and the high proportions of agency staff.

Even with a more stable organization one will have doubts whether a company uses their historical records for older projects, even if they have them!

The question is very dependent on your attitude. Some clients like to be sure that the tenderer has the necessary experience of having done similar projects a number of times before. This can produce a 'Catch 22' situation whereby new contractors are not tried out because they have not worked with the client before. This is particularly true in the public sector. Whereas other clients take the view that people get bored doing repeat work, but excited at doing new things. There is merit to both approaches.

The important part of the question is, 'What is the reputation of the company for delivering on time and within budget?' What is the point of employing an organization that consistently delivers late?! Completion of projects on time is of prime importance regardless of contract type. Whereas, completion within budget is of more relevance for reimbursable type contracts.

The reasons for any differences in performance figures are of particular importance for Fixed Price type contracts. Does the company concerned have a reputation for making claims? Will they provide a low tender price in order to be awarded the contract and then submit claims in order to recover their costs?

4.   List your experience of working in XYZ (the country of project execution).

This is another question seeking factual data. Perhaps it is of less relevance for material and equipment suppliers since they will use freight forwarding agents that specialize in the problems encountered in transport and shipping. Further, even if the vendor is involved with work on site, they will be working under the management and control of a contractor. Consequently, it is essential that a contractor has the capability and experience of dealing with cultural issues, as well as importing and exporting goods.

All clients like to see companies that show a commitment to working in their particular part of the world. A post office box address or just a sales office is hardly a sufficient demonstration of commitment to a locality.

In some cases particular location issues might be a deciding factor. For example: on a UK North Sea project it will be essential for the contractor to have an operations base in Aberdeen. Mainland European clients used to consider an office and experience on mainland Europe essential criteria.

5.   Identify your production capacity under different operating conditions.

6.   Indicate how this project will impact on your current and forecast staffing workload levels. Identify the proportion of permanent and agency staff.

You need to be confident that a supplier or contractor is operating in the middle range of their organization's resources and capability. An organization that is stretched will not be able to cope when the pressure is on or the work expands (as it does on all projects), or they take on additional work.

7.   Describe your management structure, its relationship to the ultimate parent company, and how this project fits into the organization. Describe the project management departmental structure and the different authority levels.

If you are dealing with a subsidiary company (as one usually is nowadays) it is important to know who and where decisions are made, and what the various authority levels are. Changes in senior management may mean that the company's previous experience, performance or capability is no longer valid.

This question also serves to identify the ultimate holding company that will be requested to provide a parent company guarantee when the enquiry documents are issued.

8. Provide typical Curriculum Vitae for key project positions. Provide details of work experience, education and qualifications. Indicate the status of employment of key personnel.

What are the people like? Is the company concerned composed of an experienced but aging population, or full of less experienced but enthusiastic younger people? Have they held positions of responsibility on significant projects and filled the role for meaningful periods of time? I remember analyzing one CV that at first glance looked impressive. However, the individual concerned never seemed to have started or finished anything, and never filled a role for more than a few months at a time.

The grasshopper manager, the executive who hops from job to job leaving a trail of destruction behind him, is under attack at the big oil companies.[5]

...Shell's chief executive, has imposed a job-tenure rule of four to six years for all but the most junior staff. Last week BP followed the Dutch company...In his first speech as chief executive [of BP]...a staff meeting [was told] that he wanted to stop the rapid circulation of managers, which prevented them from gaining on-the-job experience...Criticism of the rapid turnover of BP managers emerged in the Baker panel's report on the safety culture of the company's refineries in the United States after the Texas City disaster in 2005

9. Provide sample procedures and/or standards for XYZ processes.

Sample procedures for some key processes (for example, change control) provide an insight to how well a supplier's or contractor's work methods have been thought through. Are they bureaucratic? Will the consequent slowing down of decision-making impact on project completion, or are they careful to maintain the work quality and avoid rework?

10. Provide documented evidence of financial capability together with financial reports for the last 3 years. Provide a letter from a financial institution(s) that can provide references as to the company's financial capabilities.
    Alternatively, provide a letter from a financial institution willing to execute a performance bank guarantee on your behalf.

Annual reports are only 'a snapshot in time' of the financial status of a company. Consequently, it is necessary to get the accountants to carry out an analysis for trends over a period of time. However, recently established small companies may not be able to provide the run of accounts needed. Further, it is not uncommon to obtain a credit rating agency report for suppliers and manufacturers (again simply a snapshot in time) only to find them in financial difficulties during the project.

This is why it is important to have a mechanism whereby the parent company, or an insurance company, steps in and completes the contract.

11. Identify the elements of work that you propose to subcontract together with the reasons for your proposed contracting strategy.

The enquiry was issued to the chosen company, not to a whole group of subcontractors. Why have elements been subcontracted and what is the justification for it? Is it because the contractor lacks the skills or resources to do the work themselves? Further, have they the resources to manage the proposed subcontracts?

---

5   'BP halts the managerial merry-go-round to improve its safety and performance', by Carl Mortished. *The Times*, May 7, 2007.

It may be that technology transfer or involvement of local contractors is a prime objective of project execution.

12. Explain how you will execute the following key aspect to this project.

There is always a special element to a project, some issue on which the success of the project depends. Find out now if you, the client, has thought it through. Do you need to amend the enquiry document to cover something you had not thought about?

13. Identify six problem areas or issues with the execution of this project and describe how you will solve them.

This question is an 'and/or' question with number 12. I always like this question. Get six tenderers to use their best brains to solve potential problems before you start. Again, you may have to amend the enquiry document before you issue it.

There are many other questions that could be asked. However, one should consider whether they are better forming part of the enquiry document itself. For example: security measures, procedures for the sustainable use of natural resources and the protection of the environment, and safety statistics. A typical request in the enquiry document could be:

Please provide an explanation of the company's safety organization and safety procedures and how they will be used on this project. Provide details of the proposed site safety programme including training programmes, meetings and inspections. Describe the medical facilities available and proposals for site security. Detail the normal procedures for recording site safety statistics and provide actual statistics from recent projects.

However, whilst an organization's safety performance should be part of the client's database of information about suppliers and contractors, the information will not be current. A deteriorating safety record could be a good reason for a failure to pre-qualify. In these circumstances it is unreasonable to ask an organization to go through the expense and effort involved in submitting a tender.

The danger with the pre-qualification process is that once it is complete, and the chosen few have been identified, it can lead to the conclusion that any of the qualified tenderers are competent to carry out the work. This is, after all, the purpose of pre-qualification. Consequently, the lowest priced tender is the winner. This is nonsense. As a result, an evaluation process is required that avoids this pitfall.

Local purchasing

Once the main equipment and materials have been purchased, local purchasing in a foreign environment becomes much less formal. However, there is a tendency to get stuck in the formal procedural and high technology mindset. Since all projects impact on people, it is important to build relationships with the local community by making use of local ingenuity and capability.

I remember having strong arguments with the construction people who wanted to import marine plywood for use as shuttering for concrete foundations on a project in Egypt. Local contractors used any timber they could lay their hands on, since the appearance of the surface finish was not relevant once the foundations were buried.

Another argument occurred with the same people when they wanted to import steel scaffolding instead of the local custom of using bamboo. Bamboo scaffolding can be used up to amazing heights and may be safer due to its flexibility.

On a different project the construction people hired some lifting capability to string out lengths of piping similar to large tree trunks. Our safety officer was then informed that there was some equipment in use without displaying a safe working load notice, only to discover that it was the local elephant.

It can be argued that these are outside the terms of reference of a project manager. However, the project manager's job is to get the project accepted by the owner. Consequently, this may mean taking a more pragmatic or political approach to some issues, particularly if it involves getting a project finished.

For example, someone in the home office had forgotten to order the chemicals that were needed to commission and operate the water clarification and treatment plant. It was clearly going to take too long to go through the whole ordering and importation process. Fortunately, I knew that my project controls manager was, by qualification, a chemist and I gave him the task to find appropriate chemicals. He solved the problem with some innovative expertise; for example, the local custom of using crushed and baked coral to make lime (I regret that we were not sufficiently aware of the environmental impact issues at that time).

The final choice

Having pre-qualified organizations for the proposed project, a final selection needs to be made for those that will be asked to submit a tender. This selection process is carried out on the basis of the *quality* of the qualification factors, namely: experience, technical capability, resources and people, and relationships. If all things are equal and you have a surfeit of proposed tenderers, choose the ones that demonstrate that they really want the work – the hungry ones. They are the ones most likely to submit a competitive tender.

Despite the above selection and pre-qualification processes, you can still proceed on the basis of choosing a supplier that you know, like and most importantly trust. In these circumstances it is worth emphasizing, once more, that you need to be sure that you are still dealing with the same people as previously.

Again the current philosophy is to get closer to fewer suppliers. We will increase our business with you in exchange for more competitive prices. In addition, with the increased volume of business we can work together to improve productivity and reduce costs further. Eventually this may lead to a longer term partnering arrangement.

9 The Enquiry Process, Methods and Document

THE ENQUIRY PROCESS

Everything that has been covered so far has been about the whole process of buying goods and services, and there are many different expressions used to describe the process. Further, different business environments use the same three letter acronyms but apply them in a different manner to the generally accepted norm. Whilst in broad terms they are all similar, they have distinct and specific meanings and uses, as follows:

ITT: Invitation To Tender or perhaps, in view of the legal process of forming a contract, Invitation To Treat. The tender is a formal offer in response to an enquiry and can be accepted as offered. The abbreviation is also used for Instructions To Tenderers.

RFP: Request For a Proposal. This is, in effect, a request for a contractor's proposals or ideas for discussion and will result in many counter-offers before agreement is reached.

RFQ: Request For a Quotation. A quotation is usually a price for a standard catalogue item and can be accepted as offered. The abbreviation is also used for Request For Qualification Information.

ITB: Invitation To Bid. IFB: Invitation For Bidding is also used. Whilst bid and bidding are terms in common usage they have been avoided due to their use in other business contexts.

EOI: Expression Of Interest. Used to initiate an international competitive tendering process.

The UN International Trade Centre[1] has very precise meanings for RFQ, RFP and ITB as follows:

* RFQ is used for an informal invitation under US$25 000.
* RFP is used when requirements cannot be clearly or concisely defined for enquires over US$25 000.
* ITB is used when requirements are clearly and concisely defined for enquiries with a value over US$25 000.

In summary, the enquiry process consists of four distinct phases:

A. Preparation of enquiry.
B. Issue enquiry. Await tenders or proposals.
C. Evaluate tenders.
D. Negotiate and place order or contract.

---

1   See *Additional Sources* for contact details. See Chapter 11 – Additional Data 2 for the associated award criteria.

In reality, there are many more detailed steps and Figure 9.1 illustrates a generic process for the first two phases A and B.

The enquiry process will be modified to suit the choice of different tendering methods as well as the relationship of the client to the owner. A contractor acting as client for an owner will probably have to obtain their client's (the owner's) approval to various parts of the process before proceeding to the next step.

The number of steps into which the enquiry process is divided will also depend upon the size and complexity of the work involved, and the procedures of the organization concerned. Enquiries for similar categories of materials or services will follow a similar format.

The process needs a firm schedule that is developed by the project procurement manager, in conjunction with the project controls function, as early as possible in the planning process. Each enquiry should be allocated a budget and schedule. Any delay in strict adherence to the enquiry milestones could impact on the overall project schedule.

The danger is that the owner, having spent a long time evaluating options and having at last made the decision to go ahead with a particular scheme, allows too little time for the tendering process. Work performed under pressure will lead to mistakes. More importantly, believing that the excessive time taken in the early stages can be recovered, by reducing suppliers' or contractors' tender period, leads to disaster. It is in the early phases of a project that the real cost generating decisions are made, and the tenderers should be given time to exercise their expertise.

Discussions should be held with suppliers to provide advance warning and to agree reasonable periods for each stage. It should be remembered that suppliers do not have spare resources awaiting enquiries, and it may take a little time before your enquiry is entered into the supplier's system.

In today's litigious society it is important to have good administration of the enquiry process and to maintain records of the activities performed at each of the process steps. Further, the closer we get to issuing the enquiry documents the less we should talk to the proposed tenderers as individuals. All tenders should be seen to be treated equally and should receive the same information.

Notes expanding on the steps indicated in Figure 9.1 are provided below with reference to the annotated numbers in the diagram.

1.   Naturally before any work starts the project manager has a responsibility to check that the appropriate internal approval has been obtained and that the project will proceed. The project name will then be confirmed and a cost code or account number allocated.

2.   Before any enquiry can be issued an enquiry plan needs to be developed. This involves:

- Agreeing the allocation of risk in order to determine the type of contract. The type of contract will naturally impact on the time taken for each of the subsequent steps.
- Agreeing the duration of each step in the enquiry process. The time allowed will depend on the type of work (for example, supply only or design supply and erect) and the complexity of the work. The tendering period may be 3 to 6 weeks for a client risk reimbursable cost enquiry and 2 to 4 months for a contractor risk fixed price enquiry, or even longer for, say, the aerospace industry. Shorter durations will obviously be used for verbal quotations, where these are allowed by company procedures. These time periods then need to be integrated into the overall project schedule to determine the criticality of the various enquiries.
- Creating the project tender list for each of the different material and equipment types as well as the tender lists for the anticipated subcontracts.
- Deciding what elements of the enquiry process are to be electronic. For example, the Request For Information, just the pre-qualification criteria, the '[2]Request For Tender

---

2   Typical extracts from the financial press in 2006.

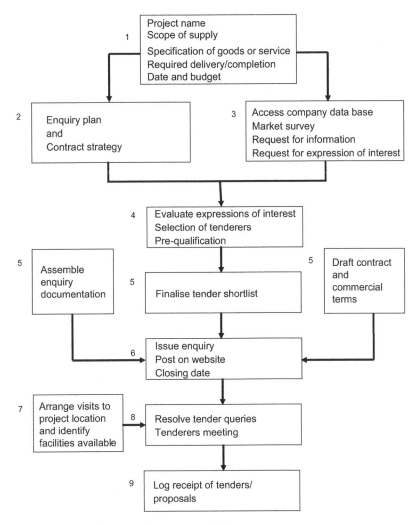

Figure 9.1   Enquiry process – phases A and B

Proposals:...the documents are available on the website,' or '[2]The complete text of the International public tender for xxx is available on [the company] website.'

- Allocating roles and responsibilities.
- Define the involvement of the owner when an organization is acting 'for and on behalf' of an owner.
- Defining the approvals required for key documents.
- The number of enquiries to be issued for each type of material, equipment and service.
- The type of enquiry and enquiry method to be used.
- Detailing the tender opening process.
- Nominating the members for the tender review board or tender evaluation team.
- Agreeing the tender analysis criteria.
- Ascertaining the authority required for placing the contract.

3.   In order to save schedule time, a client might advertise giving notice of the proposed project prior to formal approval being obtained. The notice should request interested parties to

write in and express their interest at this stage. For example, [3]'You are invited to register your interest to receive the RFP.' The advertisement or notice should provide brief details of the proposed project, its location, an indication of its size (an approximate budget value) and the areas of expertise required.

With a good corporate database it may not be necessary to survey the marketplace for companies interested in tendering. Further, with detailed and up to date information, the request for information step can be omitted. Any missing information can then be included in a pre-qualification document.

4.   If the client advertises a main contract for expressions of interest it is usual to include most, if not all, pre-qualification criteria at the same time – thus combining steps 3 and 4.

In the private sector it would be normal to select a tender list from the corporate database and have a separate pre-qualification step.

Long-lead material or equipment items must, of course, be tackled first. If their delivery period is such as to be a determining factor on the project schedule the owner may reserve manufacturing capacity, or even place orders, before awarding a contract for a main contractor. In these circumstances, the chosen main contractor will be asked to take over the orders that have already been awarded by the owner. That is the orders or contracts are novated so that they become subcontract orders to the main contract.

5.   These three activities should all occur in parallel so that by the time the tender list has been approved the enquiry can be issued.

From my perspective the tender list cannot be finalized until the project manager has signed it off as approved. It might be argued that this approval will only delay the process. However, with the right motivation and attitude they can be approved and returned to purchasing within the same day.

6.   It is advisable to use the checklist, itemized under the enquiry document later in this chapter, to make sure that all items have been included. A purchase order log should be maintained of:

- the requisition number;
- when the requisition was received;
- brief description of the materials, equipment or service required;
- the individual buyer responsible for the enquiry;
- the list of tenderers the enquiry has been sent to, and the issue date;
- the date the tenders are due to be returned;
- the tender evaluation completion date.

In many organizations these data will form part of a Material Status Report (also known as an MSR) system. This report will log the scheduled, forecast and actual dates of these data. Additional information to the above on a material status report would be:

- the selected vendor;
- the purchase order award date;
- delivery or contract completion date.

7.   Naturally, the arrangements for any visits to the project location need to be made with the local operations staff. Sufficient time needs to be set aside to enable all tenderers to visit the project location at different times. For example, for a services contract this might involve one visit in the morning and one in the afternoon for 3 days.

8.   The visit to the project location is an ideal time to resolve most queries. However, a separate tenderers' meeting (see end of Chapter 7) may also need to be arranged, again scheduled so that all tenderers are seen separately.

---

3   Typical extracts from the financial press in 2006.

As has already been discussed, it is important that all tenderers receive the same information and that all their queries are answered. Otherwise they will make assumptions that will, most likely, increase their price.

9.   It is important to have a record of the date and time of receipt of the tenders, should it be necessary to disallow any tenders due to late submission.

The more detailed steps for phases C and D are shown in Figure 9.2. and similarly expanded upon in the notes below.

10.   Details of these issues are described and discussed in Chapter 11.

11.   The timing of the notice for interviews or presentations may need to be staggered. Presentations may take place every day over a period of a week. Consequently, this needs to be planned if all tenderers are to be given the same period of time to prepare. This topic is also discussed in Chapter 11.

12.   The negotiations with any particular tenderer may not be continuous since it is important to keep at least two tenders at the same state of development.

13.   Part of the negotiations is, of course, to finalize the contract and commercial terms. However, the contracts or legal department will, as well as drafting suggested clauses for discussion in the negotiations, need to be keeping the master contract document up to date.

14.   It is important that the contract can be signed as soon as possible after everything has been finalized and agreed, and this is the purpose of the pre-award meeting. However, this cannot take place until the client's own internal approvals have been obtained.

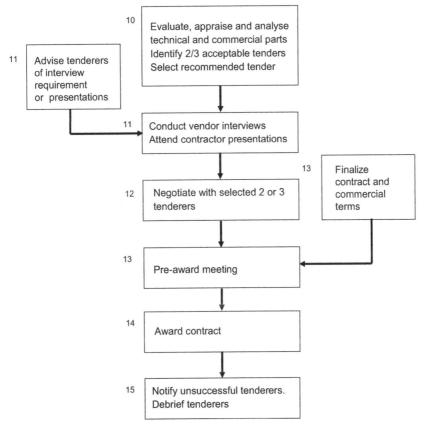

Figure 9.2   Enquiry process – phases C and D

15. Tenderers should not be kept in unnecessary suspense and should be notified that they have been unsuccessful as soon as the purchase order or contract has been finalized. For a contract for services this may be at the time of signing the contract. However, it should be noted that for a purchase order that has been sent by post, this will not be until the order has been accepted and the order acknowledgement slip returned.

A 'Practical Guide' explaining 'the contracting procedures applying to all EC external aid contracts financed from the European Communities general budget... and European Development Fund... .' is available from their website.[4]

Figure 9.3 shown below is how the generic process just described has been applied to suit the philosophy and needs of the United Nations International Trade Centre.

Finally, in order to improve the enquiry process the purchasing manager should keep track of trends in their own organization's performance as follows:

- Achievement of actual purchasing cycle milestones and overall time period compared to planned.
- Material costs compared to budget estimates should be advised to the estimating function.
- Delivery periods compared to allocated times in the project schedule should be fed back to the project controls function.

ENQUIRY METHODS

The final strategic decision to be made is the method to be used for the enquiry process, and this is now discussed in detail. Generally, competitive enquiries are sought, although this will not be the best method in all cases. In principle there are two basic methods:

1. Open tendering.
2. Selective tendering.

Other methods, described below, are really variants of selective tendering and the extreme example is non-competitive tendering by negotiation. As a generalization, competitive open tendering tends to be used more by governments and official institutions, and selective competitive tendering used more by the private sector.

Open tendering gives transparency to the procurement process so that the whole process is open to public scrutiny. It is this aspect that is important to the public sector. Whereas, selective tendering reduces the overall effort involved and the whole process can be shortened.

In the private sector the choice between open tendering and selective tendering is sometimes determined by the value of the goods or services involved. Above a certain figure open tendering is used. Similarly, below that figure selective tendering is used. The figure for materials, equipment and consultancy services will be set at one level, and contracts for design and construction services will be set at a significantly higher level.

Some clients will use a combination of open tendering and selective tendering by pre-qualifying the open tender list, in order to avoid the disadvantages of both mechanisms.

---

4 See *Additional Sources and Contact Details.*

1. Develop a plan for the enquiry process.
2. Define the basis for evaluation.
3. Specify capabilities and experience required of tenderers.
4. Prepare the enquiry package.
5. Appoint the tender or proposal opening panel.
6. Advertise the requirement.
7. Ensure suitability of those suppliers expressing their interest.
8. Issue the enquiry documents.
9. Confirm receipt by suppliers of the enquiry.
10. Clarify the enquiry document, if needed.
11. Handle requests for extensions. Only granted if enquiry amended.
12. Receive and open the responses.
13. Reject any revised offers received after the enquiry period.
14. Reject any non-compliant bids.
15. Evaluate the tenders or proposals and clarify these if necessary.
16a Undertake post-tender negotiations. (Only allowed for proposals).
16b Award the contract.
17 Debrief unsuccessful suppliers.

Figure 9.3   United Nations Trade Centre Enquiry Process[5]

Public sector procurement

In the UK the law requires all public sector procurement to follow the European Union's Directives covering procurement. The aim of the Directives is to institute a system that ensures that competition within the European Union is not distorted for public sector projects. The Directives cover; supplies, services, works and utilities. They seek to ensure transparency of the process, avoid restrictive practices and achieve the full benefits of open procurement with effective European Union wide competition. Consequently, all procurement for goods and services has to be advertised in the Official Journal of the European Union, known as the European Journal (OJEU previously known as the OJEC), and anyone is entitled to ask to be put on the tender list.

The contract value thresholds above which such contracts must be published are laid down in EU Directives and can be as low as £100 000.

The Directives dictate what, where, when and how to buy goods and services. They also cover rights and remedies for those involved. All tenderers should be treated equally and there should be no discrimination on the grounds of nationality.

That is the theory, in any case. It is not intended to go into detail here since the Office of Government Commerce (also known as the OGC) website[6] provides excellent guidelines. However, EU policy is heavily focused on the procedural process rather than the practicalities of executing projects. Further, the timescales for the various stages seem horrendously long and must contribute to the poor performance of public sector projects.

---

5   Adapted from the International Trade Centre training material 'Value in Public Procurement', 'Unit 4 The Solicitation Process', Paragraph 4.5.
6   See *Additional Sources and Contact Details*.

A cardinal rule is: do not break the rules. [7]'Holyrood building contracts violated rules says Europe.' Over the years people have got into problems not because they managed their projects badly, but because they did not follow the rules! This is true the world over.

[8]*Grinaker-LTA has launched a legal challenge over the tender process for the construction contract for the new La Mercy airport to be built north of Durban, a move that could delay the project.*

...[They] objected after its tender was excluded because it allegedly did not meet the tender requirements. ...It allegedly fell short due to its failure to put up the necessary guarantees. 'The grounds provided by the [client] for evaluating one proposal only are not justified for a project of this magnitude...and are disputed.'

...'We have written to the minister to note our dissatisfaction with the process and asked him to intervene...to ensure the correct process is followed in adjudication of the tender.'

The tender process was flawed because it did not follow the procedure outlined in the tender document, he said.

A paper[9] on 'Comparative Views of Public Procurement Reform in Gambia' sets out some useful criteria for the choice between selective and open tendering:

...procedures require that open tendering proceedings be used except when single source is required when the value of the procurement is small and competition is uneconomical. Other exceptions to the need for open tendering include situations when:

- *only one supplier can fulfil the requirement and no suitable alternative exists;*
- *there is an emergency need for the goods, works or services, involving an imminent threat to the physical safety of the population or unforeseeable urgent circumstances not due to the dilatory conduct of the procuring organization, make competition impractical;*
- *additional goods, works or services must be procured from the same source for reasons of standardization or compatibility with existing items, or;*
- *for purchase of perishable commodities purchased on competitive market terms.*

An earlier paragraph identifies some of the dilemmas of public procurement:

...Despite the potential for developing local industry through public procurement, many local and international firms do not participate in public procurement because of a perception (and at times the reality) that governments are slow payers, difficult to work with or have their own favored suppliers for contract awards. In addition to these general complaints, there is often a feeling among suppliers – based on anecdotal reports – that corruption plays a part in contract decisions.

Procurement actions should encourage suppliers to value government business and provide satisfactory quality, service and price in good time...

---

7   *The Scotsman.* 17 March, 2005.
8   Adapted from 'Grinaker-LTA to argue its case on La Mercy tender' by Samantha Ensun, *Cape Times*, January 9, 2007.
9   *Challenges in Public Procurement: Comparative Views of Public Procurement Reform in Gambia*, paper by Wayne A. Wittig and Habib Jeng. 29 August, 2004.

Open tendering

Open tendering is a procedure where all interested parties may submit tenders. In this situation the client advertises in the local, national or international press. The advertisement will provide details of the proposed project, together with an indication of the evaluation criteria, and will request interested parties to apply for the enquiry documents. In order to limit the number of parties to serious contenders, clients may ask the prospective tenderers to purchase the enquiry documents. This fee is usually non-refundable and can help offset the considerable cost of preparing the documents. Typical costs for documents in 2005 to 2007 were:

- For study consultancy work for an airport – €1000.
- An integrated hospital information system – US$2000.
- For two 1500 Megawatt nuclear reactors – €15 000.

Further, when enquiring for large international contracts it might be wise to ask the tenderers to deposit money in the form of a bank guarantee or tender bond (or bid bond) and is common practice, in some cultures. Bonds are discussed briefly at the end of Chapter 10. The hospital IT system and the two nuclear reactors, referred to above, required a 'bid security of $100 000 and a bid bond of €20 million, respectively. The bond required for the 'infrastructure construction works of 30km [of] railway line' was 'not less than 3% of the offered price.' This helps to ensure that a contract can be negotiated and signed with the successful tenderer and that they do not pull out at the last moment without justification. Re-tendering is time consuming for the client and is expensive for the tenderers. The supplier or contractor should then be asked to convert the tender bond to a performance bond for the contract. At this stage the tenderer really has no alternative but to do as requested.

The advantages are:

- Since any interested contractor or supplier is able to submit a tender it gives new or unknown parties an opportunity to compete for the work.
- It gives the appearance of the tender list being created without preconceptions.
- It ensures good competition.
- It should help prevent contractors from forming cartels.

The disadvantages are:

- Tender lists may be long, leading to excessive costs for both the client and tenderers.
- Some high quality contractors or suppliers may not submit a tender because their chances of winning are significantly reduced.
- If the lowest price is not accepted then public accountability may be questioned.
- Less able contractors or suppliers may be the ones who price on a lowest cost basis.
- If the lowest price is accepted the contract may overrun, with consequent poor quality in the execution phase.
- Prices may be depressed in the relevant business sector.

Two stage tendering

Two stage tendering can be used in open tendering or selective tendering. Not surprisingly it is a procedure where the client or procuring entity invites tenders in two stages in order to achieve a fixed price. The term is also used in different ways. In any tendering process it is important to emphasize that the better the information, for competitively tendering stage one, the more effective the process.

In the first stage a technical proposal is invited requesting complete and full information, exclusive of prices. Following a technical evaluation of all tenders received in the first stage; only those tenderers that have been found acceptable to the client are allowed to participate in the second stage. In effect, a pre-qualification process. In the second stage tenders include prices.

Alternatively, a contractor is chosen at an early stage by competitive tendering, against a notional scope of work and approximate quantities. Their method statement, the team that they offer and the resources they have available are also evaluated. The successful tenderer is then used to provide constructability into the design and develop a detailed execution plan or method statement and a programme for the work. They can also provide useful advice on which subcontractors to use, bearing in mind that they will be managing them. A contract is then negotiated for the second stage or execution phase of the work. In effect, a convertible contract or negotiated tender (see below).

Sometimes the second stage is entered into before the detail design is complete. This is not recommended since the contractor is then in a strong position to make claims.

The advantages are:

- A more collaborative arrangement is possible between the parties and relationships are developed earlier.
- The client controls the development of the design.
- The effective definition of the project in stage one provides more certainty of the total project cost.
- Earlier appointment of the main contractor to provide constructability into the design.
- Starting on site earlier and overlapping of the project phases should save time.
- There is no obligation to award a contract for the second stage.

The disadvantages are:

- Stage two is very dependent on an effective stage one.
- As the design develops additional problems or risks will be identified.
- Costs are likely to increase during the development of the design.
- The transfer of risk at stage two will increase the price.
- It requires good negotiating skills. The selected contractor has strong negotiation leverage for stage two.
- The contractor is not obliged to enter the second stage and can pull out of the negotiations.

The two envelope method

Requesting that the technical and commercial sections of a tender are submitted in separate envelopes can be known as the two envelope tendering method or process. This is not really a separate tendering method but a procedure to assist tender evaluation. The two envelope method can be used for either open tendering or selective tendering and has been used extensively in the Middle East. The concept is that the technical submission is evaluated and the commercial envelope is only opened for those that are technically acceptable. This method forces the tenderers to make sure that they conform to the enquiry. Any conditioning of the tenders is limited to requesting that the tenderer has included for the various technical requirements and that their commercial tender remains unchanged. A development of this procedure using three envelopes is identified in Chapter 11.

Selective tendering

Selective tendering, or restricted tendering (a term used by the European Union rules), means a procedure whereby only those candidates invited by the procuring entity may submit tenders.

This method is more favoured by the private sector. A short list of suppliers or contractors is usually drawn up by use of the pre-qualification process. However, the shortlist can also be selected by the client's 'tendering team' based on their experience of the suppliers' or contractors' past performances. Over the years the client will get to know the competitive, competent suppliers or contractors and this method provides them with a high level of reliability. It is better to have a limited number of really competitive tenders, all of which are likely to be technically acceptable, rather than the chance of having less competent suppliers or contractors under the open tendering method.

The advantages are:

- Since only capable contractors are invited to submit tenders the lowest *evaluated* tender can be accepted.
- The risks of failure and the cost of tendering are reduced, supposedly resulting in lower prices in the market place.
- Suppliers or contractors can include an adequate profit level and so maintain a viable industry.

The disadvantages are:

- Care needs to be taken to avoid favouritism in the selection process.
- There is a reluctance to amend lists so that the selected tenderers may get complacent. Further, new parties with new ideas may be overlooked.
- Tender prices may be higher than those from an open tender list.
- Contractors or suppliers who do not want the work may bid high, rather than declining to bid. They will do this in order to avoid getting removed from future tender or pre-qualification lists. It should be noted that, under English law, this is illegal.
- Contractors have been known to collude (forming a cartel) in order to keep prices high. Alternatively, contractors use a 'slate' system to divide up the contracts amongst themselves – also illegal.

Serial tendering

Pre-qualified contractors tender by providing a schedule of rates or a priced bill of material, together with a fixed fee for overheads and profit, for several similar projects over a period of time. In effect, a cost is agreed for each package of work or agreed project scope, as it becomes available. Alternatively, a price can be negotiated (see negotiated contracts below).

This method is more likely to be used for smaller works and limited to 2 to 3 years duration before the client re-invites tenders to validate the rates for the work.

The advantages are:

- The client and contractor can programme work over a period of time to make better use of facilities.
- It can lead to better client-contractor relationships.
- It allows a contractor more time to plan work on site and to be more efficient.
- As the client and contractor get to know each other better, and understand each other's working procedures, costs and programmes can be reduced.

The disadvantages are:

- The client might base future projects on the low cost items in the schedules.
- It reduces the work available for competition.
- The contractor may eventually become complacent.

Negotiated tenders or contracts

The client negotiates with a suitable contractor with proven experience, ideally someone that the client has worked with before. The contractor is selected early in the basic design stage on the strength of a schedule of rates. If the client has worked with them in the recent past it is easier to validate the rates quoted by the selected organization. The final scope of work (or bill of material) is drawn up when the detailed design information is available. A revised contract is then negotiated with the contractor. In effect, a special case of two stage tendering where a reimbursable contract is converted, at the appropriate time, to a fixed price' contract.

There is real merit in working with someone that you have worked with before and that you trust and like.

The advantages are:

- The client saves the time taken for the tendering process. This time saved, in getting the product into the marketplace, makes a significant impact on the business financial model. The contractor or supplier saves the tendering costs.
- The contractor or supplier can give advice to the client or client's representative at the basic design stage. Constructability is a vital input for a successful project.
- Materials for prefabrication or ordering can begin early in order to shorten the overall programme.

The disadvantages are:

- The difficulty of validating the proposed rates for the work.
- Since the client knows the details of the contractor's or supplier's pricing, it can cause difficulty when the client wants the contractor to fix their price for the agreed scope of work.
- The cost of the work could be higher since the client could be paying for the work the contractor performs to develop their fixed price.

Single source tender

If the client selects a single source tender by choice it is, in effect, a negotiated tender.

When a single source tender is forced upon the client by the circumstances of the technology or in an emergency, the client must develop a strategy to avoid being held to ransom. For example, breaking the item into smaller chunks or by redesigning the requirements so that other suppliers can be used.

Procurement routes.

The construction industry has a habit of not separating the type of contract from the tendering method (traditional or two stage tendering) or the contractual organization (management contracting and construction management). These mixed combinations are known as procurement routes. The consequence is that some combinations get entrenched in people's minds as particular types of contract – which they are not. As a result, a clear description and understanding of the allocation of risk is camouflaged. For example, the construction management route is a client managed risk contract and this should have been known from the start of the Scottish Parliament building contract.

Similarly, what I would call a scope description Design and Build (the equivalent of Engineering, Procurement and Construction or EPC) tends to be thought of as a contractor (managed) risk contract – which it can be. However, in the same way that you can have a fixed price engineering, procurement and construction contract or a reimbursable engineering, procurement and construction contract you could (although less likely in the building industry) also have a reimbursable design and build contract. In this case, the scope of the work, design and build is regarded as a procurement route rather than part of a contract strategy that could be used with open tendering, selective tendering or as a negotiated contract.

The consequence of these defined procurement routes is that other optimum contract strategies and enquiry methods are compromised or not used.

THE ENQUIRY DOCUMENT

In Chapter 4 the contract categories identified that for a fixed price contract you are buying an end product, facility or installation and the risk is managed by the supplier or contractor. Whereas, for a reimbursable contract you are buying a service and the risk remains with you, the client or owner. Consequently, the enquiry documents will need to be significantly different.

The enquiry or tender document can be one or two sheets of paper, (most likely for an enquiry for a reimbursable contract), or a compilation of a collection of documents, (more likely for an enquiry for a fixed price contract). For consultancy services type work the enquiry document is designated as the Terms Of Reference (also known as the TOR). A comprehensive invitation to tender is usually divided into separate technical and commercial sections. There is no set format for the enquiry document – although certain elements are essential – and it will typically comprise:

- Invitation transmittal letter with contact information.
- Information to tenderers containing instructions for the preparation and submission of tenders. All of the administrative matters are often included in a document entitled Instructions To Tenderers (also known as ITT and not to be confused with the invitation document!).
- A form setting out the breakdown required for the prices and the currency of the tender. This ensures that all tenders are submitted in a standard format for ease of analysis.
- Project details highlighting any unusual or non-standard features.
- A description of the facilities describing the scope of work and scope of services for which tenders are being invited.
- Job specifications and standards to be used in the design, procurement, manufacture or installation of the product.
- A coordination procedure setting out the administrative, procedural and other requirements during the execution of the project.
- The location of the work or delivery point.
- Nature of the intended contract together with the proposed terms and duties of the parties involved.
- Key project dates or programme with required delivery or completion date.
- The time and date when the quotation, tender or proposal must be submitted. The number of copies required and the required validity of the offer.

- Finally, so as to reduce the chances of a tender being incorrectly addressed, an envelope or label addressed to the client's buyer.

A number of other aspects may also need to be addressed:

- language to be used;
- confirmation of compliance with the enquiry documents;
- confidentiality statement;
- publicity approval required;
- cultural issues.

If a pre-qualification process has not been carried out then the enquiry will also include all the experience type questions discussed in Chapter 8.

In general, tenders and proposals will be written documents. However, verbal quotations are often obtained for standard catalogue items of low value. High value tenders for major works often follow a sealed tender process in order to help eliminate fraud.

It is not uncommon to request that the technical and commercial sections are submitted in separate sealed envelopes. This is administratively convenient and can facilitate clients' evaluation processes. The separate documents can then be circulated individually to the different departments. In theory, so that, for example, the technical people can perform their evaluation without being influenced by commercial matters. This eventually leads to designers with a narrow technical perspective who are not commercially minded and this makes it difficult to evaluate value for money.

The choice between the use of a simple or comprehensive invitation to tender depends on a number of factors, including:

- the type of contract intended;
- the size and scope of the project;
- the complexity of the work;
- the type of contractor to be employed;
- how the project will be controlled;
- the requirements of the financing parties involved;
- the resources available to prepare the document(s).

An international enquiry will affect the manner in which the enquiry is formulated and the terms of the contract – for example, shipping requirements.

For equipment and materials the purchasing process is initiated by a tender or quotation request. It is, in effect, a structured letter or form that can be modified to suit the specific requirements of a particular project. Most of the items identified above will be incorporated in documents forming attachments to the tender or quotation request.

A quotation request will reference the material requisition giving the technical requirements as well as the following:

- Detailed commercial requirements and terms of the proposed purchase order or contract.
- Proposed, or request for, terms of payment.
- Number of copies of invoices and packing lists.
- Timing and distribution of required documentation.
- Statement concerning the right to impose back-charges for defective items or materials.
- The applicable specification.
- A list of the standards to be used.
- Drawings and data required.

- Documents and drawings required for approval.
- Expediting, quality surveillance and inspection requirements.
- Witness test certificates required.
- Number of operating and maintenance manuals.
- Tools and recommended spares for commissioning and (usually) for 2 years operation.
- Capital spares that need to be kept in stock. For example; shafts for rotating equipment.
- Protection of openings, details of markings and packing requirements.
- Size or space limitations.
- Painting requirements.
- Transport limitations.

Equipment requests will also ask for a per diem rate to cover the cost of a service representative needed for installation and/or commissioning and start-up.

Each of the above checklist items will require more detailed explanation to ensure that there is no ambiguity in what is required of the proposed supplier. The enquiry package, compiled by the buyer specializing in the requested equipment or material, should give the supplier all the information that they need in order to provide a complete offer. Anything lacking, that means that the supplier or contractor has to make assumptions, is likely to produce higher prices or will cause more work later on clarifying the particular issue.

Care should be taken not to overburden smaller companies with bureaucracy. As a supplier said to me on one occasion, 'Oh, we always add 10 per cent when bidding to you – you ask for so much documentation.'

The enquiry package is then issued to the three or four companies on the approved tender list for the project.

The enquiry document for services may be as simple as a single page with a statement of what is intended the bidder should provide, together with six or eight questions. For example:

1. Describe the size of your company and its ability to mobilize the necessary technical and managerial expertise.
2. Provide details of projects of a similar nature that have been or are being handled by your company.
3. Describe how your company would propose to organize, staff and control this project, with special emphasis on measures to meet a compressed schedule.
4. Provide current rates for your personnel by function, including ranges for each as appropriate. If these rates include costs other than direct labour, such costs should be identified separately.
5. Provide other rate information relating to your standard personnel procedures for travel, relocation and miscellaneous expenses.
6. Identify mark-ups to direct costs for profits or fees.
7. State the general commercial terms covering the provision of your services.

At the other extreme the enquiry might consist of numerous volumes: with many detailed questions seeking to understand the procedures, processes and methods that the proposed contractor will implement on the project. Examples of more detailed questions that might be asked in the enquiry for an engineering, procurement and construction project are as follows:

1. Provide a description or method statement of how the tenderer plans to execute the various functions of engineering, procurement and construction. Identify the locations where portions of the work will be executed with a detailed description of how all of the work will be coordinated and managed.

2. Provide charts showing the proposed project organization both in the head office and branch offices and at the work location. The charts should show authority levels and lines of responsibility and communication. Nominations and alternatives should be provided for each key project position. CV's or resumes of experience, recent client references and the dates of availability should be provided for all candidates. Prior to proposing the project team the tenderer should have obtained the key team members' acceptance of their project assignment.

3. Provide an explanation of the tenderer's procurement and purchasing procedures, together with specific measures the tenderer plans to use, and an explanation of how they will be applied in the execution of this project. Provide sample project forms with an explanation of how they will be used.

This format can then be applied and modified to suit the other functions as follows:

Provide an explanation of the tenderer's:

–   scheduling organization
–   cost control organization
–   estimating methods
–   material control system
–   expediting organization
–   inspection philosophy
–   subcontracting organization
–   safety programme
–   confidentiality procedures
–   industrial relations policy

together with any techniques and how they will be applied in the execution of this project.

In most instances additional questions will need to be added in order to explore subjects in more detail. For example:

•   How will 'so and so' be monitored?
•   How will changes be handled?
•   What training is required to implement the proposed procedure?
•   How will trends be detected and control implemented?
•   Provide tabulations showing the resources available for this project.
•   Provide a projected cash flow tabulation.
•   Provide an explanation of your experience in xyz technology or technique and how you will use it to benefit this project.

The largest enquiry that I received consisted of, very conservatively, the equivalent of over 30 000 sheets of paper. This was before the days of electronic communication and the enquiry arrived as five volumes of microfilm. It consisted of all basic design drawings, data and information and specifications and standards, together with detailed questions concerning the contractors' organization, their facilities and rates for personnel, as well as a contract document. This was an enquiry for a fixed price tender and had, as near as possible, total scope definition at an advanced stage of project development, and asked all the right questions that one would ask for a reimbursable enquiry. This meant that any extras or changes would be well quantified and controlled.

Unfortunately, the enquiry is too often measured by weight, height or in feet of shelf space!

A strategic decision that should be considered on all enquiries is whether the payment terms are to be part of the competitive process. Should the payment terms be defined in the enquiry, or is the supplier or contractor to be given the opportunity to come up with their own, and possibly innovative, proposals? A low price requiring progress payments may not be cheaper than a higher price requiring payment upon completion once the Net Present Value (also known as the NPV) calculations have been carried out in order to take account of the time value of money.

A supplier or contractor may require a down payment with an order. Alternatively, they may require payment at an early stage in the contract when the client has not yet received anything of benefit to them. In these circumstances, it would be wise to request the vendor to supply an advance payment bond in return.

Finally, it is necessary to consider how inflation should be dealt with. In a low inflation marketplace the decision to request fixed prices seems fairly straightforward. Risk averse suppliers and contractors, though, are likely to add a margin, however small, above the prevailing inflation rate. Of course the longer the contract period the higher the margin is likely to be. Consequently, a reimbursable cost approach may have advantages. It may be that, despite a low inflation economy, fuel prices or particular materials such as copper are subject to price fluctuations that a supplier or contractor finds unacceptable. As a client one should be aware of this possibility and decide whether to include a Contract Price Adjustment, also known as a CPA, clause in the enquiry. The inclusion of such a clause transfers the inflation risk to the client and may stop the tenderer from carrying any of the associated risk themselves.

The importance of the quality of the enquiry document cannot be overstated. If the invitation to tender has been well prepared it provides a sound basis for obtaining and evaluating competitive offers.

10 *Payment Strategies*

The purchaser has two choices. They can either dictate the commercial terms or let them be part of the competitive tendering process. If the tenderer is given the opportunity they will want regular payment at frequent intervals in order to spread their risk; ideally, as much money as possible, as early as possible. Whereas, I have no doubts that the purchaser should not pay anything until they have received something of value. This may mean that the supplier is not paid anything until the order or contract is complete – 100 per cent upon completion. The immediate response to this situation is that the supplier or contractor complains that they have to finance the project. There is then a negotiation to find something acceptable to both parties.

The two options of 100 per cent payment upon completion, or stage payments at agreed intervals, can be applied to all the different types of contract. There can be a fixed price contract with 100 per cent payment upon completion or with payments at agreed stages. Remeasured or rates type contracts are most likely to be paid at monthly intervals. Again a reimbursable cost contract may be paid at monthly intervals or as and when the costs are incurred. On the other hand, the client may only reimburse the costs when agreed targets have been met. These terms demonstrate that just using the broad expressions fixed price and reimbursable only describes part of the risk allocation. The full payment terms need to be detailed to correctly describe the type of contract within a risk category. The shorter the payment time scale, the more risk is assumed or allocated to the client. Thus, as one moves down the list below the risk to the client increases.

- Fixed price   100 per cent payment on meeting performance criteria.
- Fixed price   100 per cent payment upon completion.
- Fixed price   with stage payments.
- Reimbursable   payment on completion of targets.
- Reimbursable   monthly payments.
- Reimbursable   payments made as and when costs are incurred.

Having negotiated and signed a contract, the contract creates an obligation in law to pay, if the contractor performs. However, there is probably a lack of trust between the parties. The purchaser does not know if the supplier will perform and the seller does not know if the client will pay them. Thus, there has to be a payment process agreed between the parties, and the supplier has to deliver the product or perform the service on time and in accordance with the specification. Correctly structured, the payment terms should act as a motivator to the supplier or contractor to achieve the project objectives. Consequently, issues that need to be resolved are:

- How will payment take place, what process will be used?
- Where should delivery take place?

- What should be supplied for a payment to be made?
- When should the supplier be paid?
- How or in what form, and in what currency, should the supplier be paid?
- How long will the payment process take?
- How can the buyer and seller be satisfied that their interests are protected?

TERMS OF PAYMENT

The terms of payment for goods and services performed in the same country are significantly more straightforward than when contracting between different countries. When contracting between countries for the delivery of goods, you must decide when the liability for delivery changes from the supplier to the buyer. Thus, where delivery should take place is determined by the Incoterms, and these were described in Chapter 5 which dealt with contracting issues.

It is also more straightforward to define the terms of payment for a contractor managed risk (fixed price) contract, since only the sums to be paid and the agreed intervals have to be defined. Since the intent of the contract is for the supplier or contractor to deliver an end product, then payment should only be made for the value of an item when it is delivered. However, there must be sufficient money left in the contract to motivate the supplier or contractor to complete any remaining small value items that are part of the contract.

As indicated above, payment should only be made when something of value is supplied. With progress payments, the difficulty is balancing payment schedules with the work performed. A contractor will want to, and will try to, front-end load their payment terms by placing a disproportionate overhead and profit element on early activities. For example, setting up or site establishment, site clearance, excavation activities and buying and delivering materials. This subject is discussed in detail in Chapter 12. Performing a discounted cash flow calculation for front-end loaded payments produces a significant advantage for a contractor's project profit element. A retention helps: to guard against overpayment, to allow for errors, to enable faulty work to be put right and to inject some (although minimal) motivation.

The problem of overpayment is less likely to occur with shared risk contract forms involving schedules and rates, since regular measurements (usually monthly) need to be made to determine the value of the work performed.

If the client is controlling the risk and is reimbursing costs, considerably more checking will be required. As well as accounting procedures, the client needs to check the contractor's procedures for:

- Labour:
  - Is there a need for it?
  - How much and on what basis?
- Materials:
  - Is the contractor's buying capability satisfactory?
  - Do the goods delivered need to be checked?
- Plant:
  - Is the plant requested necessary?
  - How is it to be paid for?
  - Will it be economic to buy and resell or should the plant be hired?
  - Will the plant have any residual value?

Payment procedure

Payment is dependent upon making progress. However, the timing of the payment depends upon a number of other events happening at roughly the same time. In sequence they are:

- Evidence of achieving a payment stage: If evidence has been agreed, say drawings, it is prudent to check that they are of an appropriate standard.
- Availability of project funds: The initiative should be taken by the client and the contractor should be asked to supply a projected cash flow forecast, which should be included as part of the enquiry process. However, if the contractor wants to be paid promptly, they should provide advance warning that they anticipate achieving a payment milestone.
- Invoice submitted on time: Too often the supplier or contractor does not submit the invoice until the due date and does not allow time for the client's payment procedures – but this is the supplier's or contractor's problem.
- Invoice correct: Often minor errors can be used as an excuse for delaying payment. The invoice should still be paid but minus any disputed amounts.

Any delay in any one of these stages will mean that the whole process will be started again, usually to the detriment of the supplier or contractor rather than the client.

Payment stages

As indicated at the beginning of this chapter the following options are all possible payment stages:

- When the order is placed – 'with order'.
- Monthly or according to agreed stages or schedule.
- Upon achieving an agreed milestone.
- When the goods are delivered or the project is complete – '100 per cent upon completion'.

It should be emphasized that the earlier the payment is made the higher the risk to the client.

Payments with order

It should be obvious that a 100 per cent payment with an order should never be used. However, in the mid 1990s, GEC received just this for a power station contract. It would have been interesting to listen to the discussions at the first meeting arguing about the lack of progress.

Any early payment must be for a justifiable reason. However, the payment needs careful structuring to prevent a supplier or contractor avoiding their contractual obligations. There is also a danger of overpayment for the goods or services provided. In civil engineering and building projects it is common to request an early payment for site establishment and/or purchase of materials. Once the materials have been paid for there is no longer an incentive to look after them and degradation will occur. The supplier or contractor can then claim for additional work to replace lost materials or repair of damaged equipment. Consequently, a contractor should be asked to purchase materials in their own name. Any payment reimbursing the costs, for materials and equipment, should not be made until they have been inspected as conforming to the specification and have been incorporated into the finished facility.

Stage payments

Stage payments appear to be an equitable arrangement and this is optimal when the client and contractor have a neutral cash flow situation. The contractor gets paid for work done as and when it is completed. Usually the payment intervals are at agreed 'milestones'. There are, however, some significant problems:

i.   The difficulty of defining when a milestone has been achieved or completed.

ii.   As already indicated, if the payments are badly 'weighted' there is a diminishing incentive for the supplier or contractor to complete the order or the project.

The problem for subcontractors is that they will probably have the same terms imposed upon them that the main contractor (their client) has negotiated with the owner client. The commercial situation used to be worse, in that the contractor client only paid the subcontractor when they had received payment from the owner client. Since the Latham report1 'pay when paid' has become illegal.

100 per cent payment upon completion

This is the ideal from the client's perspective. It is the natural choice for the supply of materials and equipment. It provides the client with the maximum leverage to get the materials and equipment delivered, or the project completed, and ensures that there are no loose ends. These terms should always be used for a performance specified contract. If the project, piece of equipment or system does not work then the client does not pay for it and, consequently, does not lose money (apart from the not insignificant benefit to be obtained from the project). If this approach had been used for the Stock Exchange (computerization of deals) Taurus Project then the client would have saved themselves £300m.

The contractor, on the other hand, will complain bitterly that they are bearing all the financing costs. This is true, but it is doubtful if they have taken the trouble to calculate the actual figures. Further, in a competitive tendering situation the tenderer may not include all these costs in their offer. On the other hand, with stage payments the client is losing 100 per cent of the 'value' of money as soon as it is paid – when it has been paid, it is not earning. On this basis it is preferable to get the supplier or contractor to bear the financing costs and maintain the financial motivation on the supplier or contractor.

The following is an example of payment terms for a large piece of equipment. In reality the equipment was sufficiently large that it had to be lifted over obstructions on site, thus requiring a special crane. It was sufficiently special that if the booked time and date were missed, the crane would not be available for another 3 months. Consequently, it was decided that the delivery risk was best managed by the supplier of the equipment, and the terms were written as 'delivered onto foundations'. Other issues led the project team, to choose 100 per cent payment on completion of order, as payment terms. During the tender process the suppliers suggested that it was more appropriate for the construction people on site to be responsible for the lifting procedure. There was also the argument about financing costs as described above, but the temptation to change the strategy was resisted. The equipment was delivered onto its foundations without problems and an invoice issued for the £5m (early 1980s money). The buyer concerned pointed out that payment was not due until the order was complete. This went on for some time despite emphasizing to the vendor that the second item on the order had not been provided; only to be met with, 'But we have delivered the equipment.' Eventually, the procurement people asked me, as project manager, to get involved. I remember quite vividly that the managing director of the company concerned was confrontational and aggressive, if not insulting. Eventually he realized that, despite having delivered a £5m piece of equipment, we were not going to pay until Item 2 on the order was provided. Item 2 being various items of documentation that we needed to complete our contractual obligations.

---

1   'Constructing The Team' by Sir Michael Latham. A report for the Government into the construction industry's shortcomings. HMSO, 1994.

When the MD realized that £5m depended on pieces of paper, our documentation was received fairly promptly!

If we had made progress payments, such that there was only 10 per cent remaining for finalizing the last elements of the order, I doubt if we would have had the same attention. This scenario was revalidated for me when in 2003 I performed some work for a major UK rotating machinery manufacturer. They admitted to me that they had not received payment for an order, because they had not delivered the operating and maintenance manuals, a year after the equipment had been delivered!

Payment for what?

Payment can be made for anything agreed by the parties to the contract. Common items in order of project timing sequence are:

- Issue of drawings or other documents.
- Delivery of goods in accordance with the agreed Incoterms.
- Substantial/mechanical completion or beneficial occupation as defined in the contract.
- Project completion (including documentation).
- Three months after achieving project performance criteria.

Again it should be noted that the earlier the payment is made the higher the risk to the client.

If goods are not delivered to the client's premises or site, where they can be examined, there is the difficulty of proof that the goods have been delivered to the agreed intermediate location. If the goods are not physically examined then the proof of delivery is provided by documentation.

Typical documentation that is used to demonstrate that the seller has performed specific actions is listed below. Some of these may be required in combination in order to trigger the payment process.

- certificate of origin
- Bill of Lading
- air way bill
- warehouse receipt
- forwarders receipt
- inspection certificate
- customs submissions
- insurance documents
- certificate of conformity
- certificate of (customer) acceptance
- invoices

The documents must clearly state the criteria that will be acceptable to enable a bank to make payment, when the documents are submitted; for example, a 'clean' bill of lading. Any annotations along the lines of: 'crate damaged' on any of these documents are likely to lead to a bank refusing to make payment to a supplier, thus protecting the client's interests.

How long for the payment process?

Typical durations for the payment process to take place are:

- Net monthly account.
- Net cash 30 days/60 days.

- End of month following month of invoice issue/receipt.

Rather than have a different number of days for each month, net 30 days is used. It is used in accounting practice to help comparison of monthly averages of certain parameters.

Professional clients should set up 'back to back' bank accounts so that payment funds are transferred from the client's project account, direct to the contractor's project account, as soon as payment is authorized.

However, it is not unknown for clients to slow down the payment in a mistaken belief that they are motivating the contractor. Sometimes this is done in order to improve their own cash flow. For example, 60 days after the end of the month in which the invoice is received: this could be 60 days or even 90 days if the invoice is received at the beginning of a month. This penalizing of the contractor slows down the work and the client suffers in the long term.

One thing that must be avoided is overpaying the contractor, which is very common in the domestic building renovation market. The contractor then disappears to try the whole process over again with another client.

How payment is to be made

Payment can be made by any of the following mechanisms:

- open account
    - cash or cheque
    - bank transfer – by BACS or CHAPS
    - bill of exchange
- letter of credit

Unfortunately, from the seller's point of view and until trust has been established between the parties, there are problems with some of these mechanisms.

Open account is when the buyer pays the seller for the goods on presentation of an invoice or other documents that they agree to, for example, an inspection certificate. The buyer might pay with cash, but from the sellers point of view the cash might be forged and, in any case, it is difficult to transfer. Money laundering regulations make this option more trouble than it is worth. Similarly, there may be insufficient funds to cover a cheque and it will then be 'returned to drawer'.

A bank transfer, on the other hand, involves a third party that has satisfied themselves that funds are available. BACS (Bank Automated Credit System – being phased out) is a 3-day transfer and CHAPS (Clearing House Automated Payment System) is an electronic same day transfer of funds. The bank transfer option will be used when the client has satisfied themselves that a payment is due and they instruct the bank accordingly. Whereas a bank transfer against documents will be used when a client has established the payment procedure and defined the documents required to prove that the supplier or contractor has performed. The bank will make payment (provided they have satisfied themselves that the documents are correct), without requiring a direct instruction from the purchaser. This option will most likely be used with a letter of credit – see below.

A bill of exchange is a type of cheque or promissory note without interest, in effect an IOU. It must be in writing and signed and dated. It is used primarily in international trade, and is a written unconditional order by a person or business which directs the recipient to pay a fixed sum of money to a third party at a future date. If the bill of exchange is drawn on a bank, it is called a bank draft. If it is drawn on another party, it is called a trade draft. Sometimes a bill of exchange will simply be called a draft but, whereas a draft is always negotiable (transferable by endorsement), this is not necessarily true of a bill of exchange.

Countertrade

The consideration required to form a contract must be something of value. It need not be money. This is the basis of countertrade. The consideration is in the form of goods and services.

[2]*When governments do international trade deals, what they buy is not always the chief consideration. As a recent US Department of Commerce report to Congress explained, 'Some governments readily admit that they are no longer concerned with the price or the quality of the defence system purchased, but rather with the scope of the offset package offered.'*

'Recently, the Czech Republic announced that in competition for its jet fighter procurement, offset would be the deciding factor as opposed to technical and performance criteria and price.' Such nuggets help explain why interest is taking off in offset – or countertrade – as this form of finance through reciprocal trade is also known.

According to Trade Partners UK, a UK government agency that promotes British companies abroad, between 5 per cent and 40 per cent of world trade is countertrade related, an indication of how substantial a part it now plays in international political and commercial transactions.

Countertrade is also growing. For example, it is spreading from arms and aerospace purchases to civil infrastructure projects…

There are five main strands to countertrade: barter; buy-back; counterpurchase; tolling; and offset, the biggest of the lot, which in turn divides into direct and indirect offsets. Two or more of the five may be stitched together to create a countertrade and if that is not complicated enough the terms are sometimes used interchangeably.

In direct offset, the supplier agrees to incorporate materials, components or sub-assemblies procured from the importer country. It is a way of promoting import substitution and local manufacture of defence or other equipment.

With indirect offset, suppliers enter into long-term cooperation and undertake to stimulate inward investment unconnected to the supply contract – again to support, increase or diversify the local industrial base.

These investments need not be made by the company winning the original contract and may be totally different in nature from the first sale…

'Many developing countries find it hard to buy much-needed goods or equipment due to lack of foreign exchange,' …In some countries, countertrade may be the only effective mechanism for doing business.

TERMS OF PAYMENT CASE[3]

The following exercise is designed to illustrate that the payment terms are an integral part of the contracting strategy.

You represent a major corporation that wants to set up a 2-year master services agreement with Cushy Consultants Ltd. Your requirements are for the occasional services of some specialists for periods of 2 to 3 weeks at a time. The market demand for the specialists is very high and their pay rates are moving up very quickly. What contract payment terms are appropriate?

---

2   'An international deal with strings attached' by Ross Davis. *Financial Times*, July 18, 2002.

3   When, as ,manager of proposals, we were asked to tender for this work I declined, since I decided there were too many risks involved. I did not think through the issues and the solutions that would make it a workable proposition.

The development of a suggested solution to the issues involved is described below:

Supplier – Cushy Consultants:

'We require £X per hour with an escalation clause (based on named industry norms) revising the rates every 3 months.'

Buyer – your corporation:

'Fine. I would like six of your best (or the following named) people for Monday.'

Supplier:

'Because we may have to hire additional agency people to cover for our own work (we are very busy) we will need a minimum of 2 weeks notice and we choose the people.'

Buyer:

'But you may not supply people who we regard as competent enough. Let us agree a "pool of people", who are acceptable, and you nominate from the pool. '

Supplier:

'Agreed. By the way some of them have just been promoted, so their rates have increased!'

Buyer:

'No promotions allowed during the course of the work. Provided that we agree that the individuals concerned have demonstrated superior performance, we will consider a revised rate at the next joint review date.'

After 3 days you tell the supplier that your work requirements have changed. In addition, you no longer require the services of three of the people for a further 10 days.

Note: The supplier has hired additional people and is involved in more administration and costs than they had allowed for.

Supplier:

'We need a retainer with a minimum hire period and 2 weeks notice of termination.'

Has this covered all the problems that are likely to occur? Is it now a workable agreement?

PAYMENT AND CONTRACT SECURITY

The purchaser requires security for any advance or progress payments, in case the supplier or contractor does not perform, and the supplier or contractor requires some security that payment will be made. Furthermore, a client failing to pay promptly creates a lack of trust and, consequently, reinforces the desire for security of payment when the supplier or contractor does perform as requested. This security is usually provided through an independent third party – a bank. A bank will provide what are termed 'documentary credits'. A documentary credit is, 'A written undertaking given by a bank on behalf of the buyer, to pay the seller an amount of money within a specified time, provided the seller presents documents strictly in accordance with the terms laid down in the letter of credit.'

It is, therefore, important to understand that banks deal with documents, and not the goods that the documents represent, and that the documents must strictly comply with the credit terms in the letter.

Letters of credit

A letter of credit is literally a letter, usually conditional, saying that the purchaser's bank will make payment when the bank receives documents providing evidence that certain agreed criteria have been met.

The documentation required can comprise any of those listed earlier under 'payment for what?' For example, invoice, bill of lading, material receipt records and inspection certificate. There are three types of letter:

- revocable
- irrevocable
- confirmed

From the seller's point of view the revocable letter of credit, one that can be cancelled at any time (for example, just as the ship has sailed with the seller's goods), is not worth the paper it is written on. The irrevocable letter of credit is only slightly better in that it cannot be cancelled. However, it will have been issued by a foreign bank in a foreign country, and the seller may not be certain about the viability or integrity of the issuing organization. Further, the buyer's relationship with the local financial organization may enable them to exert pressure (for example, a foreign government changing the local law) in order to delay payment. Consequently, the only letter of credit that, from the seller's point of view, has any meaningful substance to it is 'a confirmed irrevocable letter of credit'. In this situation the purchaser's bank (the issuing bank) agrees to reimburse a bank nominated by the seller (the advising bank) provided they are satisfied with the documentation.

A confirmed irrevocable letter of credit should be as solidly reliable as a gold bar! However, on a project in Sri Lanka, the Minister of Finance boldly stated in Parliament (when equipment deliveries were in full swing), 'This project is not financially viable.' Accordingly, all the (rather arrogantly termed) first class western banks froze their irrevocable confirmed letters of credit. The consequence of this was that one enterprizing seller managed to have a ship arrested on the high seas in order to repossess their goods! If this can be done once it can be done again. Thus, are confirmed irrevocable letters of credit really as good as gold?

Guarantees and bonds

Guarantees are necessary to ensure that contracts can be completed in the event of the failure of the supplier or contractor. Guarantees can either be issued by a bank or the parent company of the organization with which one is contracting. Guarantees take two forms; they either provide funds or undertake to get any remaining work completed. In general, guarantees issued by banks provide for the payment of costs in the event of default of the supplier or contractor. Whereas, parent company guarantees are for the entire performance of the contract, without limitation. A common feature of guarantees is that the parties are jointly and severally liable. This gives the client the option to claim against either the guarantor or the contractor.

The wording on these documents is critical. There have been a number of examples where the intended results were not achieved because of a failure in the wording of the documents. Therefore, always employ someone who specializes in this type of work.

In a number of countries insurance companies, rather than banks, provide the guarantee. Unless the requirement for guarantees has been advised at an early stage, contract negotiations can be slowed down. This is due to the need for the insurance company to investigate the tenderer who needs to provide the guarantee.

In North America guarantees are provided by surety companies since banks are prohibited from issuing guarantees. The surety company will arrange for a contract to be completed by another contractor in the event of default. As a consequence, they may get heavily involved in the contract negotiations with the client and even get involved in the execution of the contract, as well as monitoring the client's performance.

In the past, the use of financial guarantees in the UK has been somewhat limited since contractual remedies can be enforced through the legal system. Further, the theory was that the financial status of a company could be reasonably well established. More recent high profile company collapses have called this into question and their use should be considered more often.

Bonds

Guarantees issued by banks commonly take the form of a bond. A bond is best described as a large bank note. Lord Denning, when he was Master of the Rolls, likened them to promissory notes. Naturally, the issuing bank will require the seller to provide assets, as security, to the same value as the bond. Unlike a letter of credit, a bond need not be conditional on evidence being provided for the funds it represents to be drawn upon. The advantage for the client is that they do not have to prove that the contractor has defaulted. If a bond is conditional then, once all the conditions have been met, they become 'on demand'. A bond is an entirely separate obligation from the contract to which it is associated and cannot be withdrawn by the seller.

It is normal practice in the mechanical and petrochemical industries to request 'on demand' bonds. Whereas, in the building industry, 'conditional bonds' are normal practice. Typical project bonds are:

- tender or bid bond
- advance payment bond
- performance bond
- retention bond
- subcontractors' bond

Tender bond

A tender bond is necessary to ensure that a company can be held to their tender once it has been submitted. It helps to eliminate companies who do not have the financial resources to execute the proposed contract. As a result, it is most likely to be used in open tendering situations since pre-qualification should have resolved the financial issues prior to issuing the enquiry documents.

An example from the press: In October 2006, the construction of a public building requested a 'bank guarantee/participation bond which must not amount to less than 3 per cent of the total value of the procurement.'

Advance payment bond

An advance payment bond is used, as its name implies, when a contractor has requested a mobilization or site establishment payment as a pre-condition to starting work. It should help to reduce the overpayment situation described earlier. However, the bond may be worded in such a way that its value decreases as work progresses.

Performance bond

The performance bond enables the client to impose a financial penalty in the event that the supplier or contractor fails to perform.

Retention bond

A retention bond can be used instead of retention monies. This is attractive to a supplier or contractor since it enables them to be paid the retention cash that would otherwise be withheld during the maintenance period. It should only be considered once the contract is completed.

Subcontractor bond

Subcontractor bonds can be requested for a proportion of the value of the subcontracts. They are used to enable the client to pay the subcontractors in the event that the contractor goes into liquidation.

All of the issues in this chapter are designed so that a negotiated solution can be developed that meets the needs of the contracting parties.

Whilst the mechanisms described provide objective criteria for both parties, it must be emphasized that precision in the formulation and wording of the documentation is vital. Use specialists, expert in the specific areas concerned.

11 *Evaluating the Tenders*

This is a crucial stage in the development of a project since by the end of this stage, in the purchase of goods or services, the management initiative passes from the buyer to the seller. Further, tender evaluation time is a constant source of frustration time to project managers – often having a direct impact on the project completion date.

The opening of sealed tenders can be a very private activity with only those with a 'need to know' attending. Alternatively, at the other extreme, it can be very public (usually for large international tenders) with all tenderers having a representative to see who was cleverest at presenting the price to make it appear to be the lowest. I have attended a public opening of tenders only to see the final award order to be the reverse of the order recorded at the opening!

The first step in the evaluation process is to record the tenders received. Tenders or proposals should be opened using arrangements designed to prevent malpractice. An opening panel or committee is essential, even for small purchases, if matters are to be seen to be done correctly. The tendering plan should have determined who would be part of a formal tender opening process. As a consequence tenders should be placed in a secure facility until the tender opening committee is convened, when all tenders have been received. Tenders for smaller works or goods and materials may be opened, as and when received, by a project panel.

The tender opening panel can be composed solely of project representatives, for example, project manager, project procurement manager and project engineer. However, in order to prevent malpractice some organizations operate with two more formal committees. A Technical Tender Opening Committee and a Commercial Opening Committee; comprising corporate heads of departments attending on a rotating basis. The technical committee meets after hours and opens the technical tenders, the first envelope, from all the different projects. The tenders are checked for completeness and compliance with the enquiry requirements. A check is carried out to confirm that missing documents, say the tender bond, are in the sealed commercial envelope. The technical packages are then sent for technical evaluation. Upon completion of the technical evaluation the tenderers are instructed to submit a third sealed envelope identifying the price impacts of the technical conditioning to their original offer. The commercial tender opening committee – chaired by the corporate manager of procurement – then opens the technically acceptable commercial tenders comprising the second envelopes together with the third envelopes, in order to identify the lowest technically acceptable price.

On the other hand, it is sensible (in order to save time and unnecessary work) to carry out a preliminary assessment by comparing the various elements of the tenders against the purchaser's own budget estimate. This identifies any tenders that are clearly not going to be acceptable because they are significantly in excess of the budget, or have no commercial merit. After management approval they can be dropped from being evaluated and advised that they have not been successful.

The remaining technical tenders are then assessed, on a comparable basis with any questions, which are required to make the tender compliant with the enquiry, numbered and logged.

An order of merit is determined and the tenderers are asked to submit the cost impact of each numbered query and their revised tender price. This provides an audit trail as to how the final price has been reached, and also indicates what the technical conditioning has done to the price.

Final post-tender negotiations then take place to eliminate queries, minor commercial or technical omissions or qualifications.

If all tenders are significantly different from the purchaser's budget estimate then the purchaser should question whether their specification or scope was sufficiently clear.

The analysis seeks to determine the best combination of project execution factors and business terms. The difficult part is selecting the assessment process and evaluation criteria to be used. The European Journal requires public sector projects to declare the evaluation criteria. However, the weighting or points awarded to the various elements does not have to be identified. Additional Data 1 to 3 (at the end of this chapter) provide a variety of scoring criteria from different sources, and across different business sectors.

SELECTION AND EVALUATION PROCESSES

The selection process should form part of the overall contracting strategy and the method should be chosen on the basis that it will help to achieve the owner's or client's objectives. The difficulty in deciding on the evaluation criteria is described in a paper I wrote in 1983[1].

In making the final choice during the procurement of equipment, three fundamental and distinct criteria need to be examined.

- *Functional parameters.*
- *Schedule and cost requirements.*
- *Operating costs.*
  - *Does the equipment meet the functional parameters?*
  - *Does the equipment meet the schedule and cost requirements for mechanical completion of the plant?*
  - *Does the equipment meet the operating costs required for the plant?*

Delineation of the responsibility is dictated by the emphasis forced on each of these three factors. It could be said that:

- *The manufacturer's prime interest is in the first requirement.*
- *The engineering contractor's interest is in the second requirement.*
- *The user's prime interest is in the third.*

The cheapest engineered machine is one that will do the job required with no concern for the time required for erection and commissioning, or for the cost of operation. It is because of this that there is a need to impose contractual specifications in order to eliminate all the little details that cause delays on site.

The cheapest machine is almost always the machine with the lowest operating efficiency and highest maintenance cost. The user's need to reduce operating costs imposes further specifications to ensure

---

1   'The Engineering Function in Petroleum and Gas Projects', paper presented by G.G.F. Ward to the Federation of European Petroleum and Gas Equipment Manufacturers Conference October, 1983.

conservative design. Attention to operability and maintenance requirements leads to the addition of further specifications.

As a generalization, we all like a bargain and are attracted by something for nothing. Why else does 'buy one get one free' or '2 for 1' work so well? On the other hand, people are willing to pay for quality, and even for perceived quality. How else could one sell a product for 50 pounds an ounce that costs 50 pence to make, namely perfume?

In the process and power industries it is recognized that the right quality will cost more, and in aerospace and the nuclear industries quality is paramount. However, in the construction and building industry there seems to be a belief that the lowest price is offering the same quality as the highest price. The ultimate proof of the flawed belief that the highest quality can be obtained for the lowest price is, of course, the Scottish Parliament fiasco. In the rail industry there are different driving forces:

[2]Railways are inherently expensive, and most routes are, in effect, supported by the government, ...

...Prices are not rising because operators are buying fancy new trains but because the government is seeking to claw back more of its subsidy. Companies who win the bidding race to run lines are those that promise to return more money to the government. ...Passengers will continue to pay more as long as the government thinks rail users are the only beneficiaries of the train... Other European governments recognise the broader economic benefits of rail travel – relieving congestion, regenerating rural areas – as well as trains being safer than cars.

In 1850 John Ruskin said, 'There is hardly anything in the world that some man cannot make a little worse and sell a little cheaper, and the people who consider price only are this man's lawful prey.'

Consequently, as well as the money aspects, one needs to analyze their technical acceptability and their commercial viability and, of course, their ability to meet the delivery dates or contract schedule. All criteria for carrying out tender appraisals will impact on each other. Thus the appraisal must be carried out as a client team effort. The evaluation team must be composed of people who have sufficient experience to cover:

- technical aspects
- execution aspects
  - design
  - procurement
  - installation/construction
  - implementation
  - project controls
- commercial issues
- contractual and legal aspects

Each of the review team members should have one topic to assess all the way through the evaluation phase. There may be a number of specialists involved for the technical aspects and, ideally, (for large contracts) each of the main topic areas should be assessed by more than one person.

Within contractors it is not uncommon to restrict engineers to purely technical matters. Not surprisingly this develops an organization that is commercially naive.

---

2   'Are train fares too high?' by Patrick Barkham. *The Guardian*, 4 January, 2006

It is vital that the evaluation process is carried out as objectively as possible and confidentiality maintained. Further, it will be necessary to determine how, and if, one could work with any one of the tenderers, even the least preferred option. In the end, the remaining evaluation criteria may, on balance, override a deficiency in one particular area, and one should be prepared for this possibility.

There are a surprising number of selection philosophies, probably because all evaluation processes involve significant amounts of subjectivity. The subjectivity involved means that most evaluation criteria can be manipulated so that the evaluator can, if they so wish, choose who they want. It is interesting that, on the often used example of the Scottish Parliament building, the highest priced tender (which was originally dropped from the evaluation) was awarded the contract.

A number of evaluation processes are included for the sake of completeness rather than as recommendations.

a) Choosing whom you want may be the least complicated method, and it may also be the most sophisticated method.

Select someone capable with a demonstrated record of designing to a budget and meeting schedules. When I joined Bechtel they did not have any work that had been tendered for competitively. All their projects had been given to them because of their reputation of, not only completing projects (financing organizations like this), but completing them on, or ahead, of schedule and under budget. I read in the press (prior to the award of the Channel Tunnel Rail Link contract) that the Deputy Prime Minister had said that, 'If you want to do infrastructure projects in this country, you had better have Bechtel on your team.'[3] This option is, of course, used in conjunction with the next item – b).

b) Negotiate with a preferred tenderer. There should be a logical reason for the choice of the preferred bidder – as in a) above. A client that carries out many similar projects should know the key parameters of their projects and the marketplace. As a result, they should be able to negotiate with a selected contractor, with a sufficient degree of confidence and, hence, benefit from reduced tendering and evaluation time periods. However, the danger is that the preferred tenderer takes advantage of their vastly enhanced negotiation leverage. This seems to be the problem that one often reads about in the press when the government cannot do a deal with the chosen contractor and have had to choose a new preferred tenderer – usually for their larger projects. The significant loss of schedule time (by negotiating in series) negates all of the benefits of the negotiation process. It is, therefore, essential to keep at least two tenderers under consideration and negotiate with them in parallel.

c) Probably the most common method; choose the lowest price from well recognized brands (without any real evaluation). For example, who got fired for saving money and choosing say IBM, HP or Cannon? And there are equally recognizable names in all business environments. This lowest price tactic is often driven by 'the fear of the auditor syndrome'. The approach is illogical, since the amount of money saved is not significant in the overall cost of a project. However, the time and schedule lost in rectifying the problems caused by the wrong low tender can be substantial.

d) The two envelope tendering process. All tenders are checked for technical acceptability (the first envelope). The second envelope, containing the price, is only opened for those tenders that are technically acceptable, and the lowest price is selected. The unsuccessful tenderers then have their second envelope returned unopened. That is the theory. It is regarded as unethical to try and improve ones estimating database with a quick peek at the remaining tenders.

---

3 Source unknown by author.

e)   Throw away the lowest price – they have made the most mistakes. Throw away the highest price – they are clearly not interested. Award the contract to the average of the remainder – probably the true cost of the job. For interest, further options are shown in Additional Data 1 at the end of this chapter.

f)   Value for money. This is frequently used for UK Government goods and services. However, it is extremely difficult to define 'value for money' and the concept can, in effect, be manipulated to mean whatever the evaluator wants it to mean. Thus, the subjectivity of the process can often be the focus of political criticism.

g)   Use a point scoring system to evaluate the tenders. A set of criteria are chosen and marks are awarded out of a total weighting given to each of the criteria. The problem with this procedure is in the selection of the criteria to be used and the weighting of the points for the scoring process. Additional Data 2 at the end of this chapter provides an example of such a system. An example in Additional Data 1 endeavours to combine a points scoring system with quoted prices.

Banks letting international consultancy work tend to declare the details of the evaluation criteria and point count system. The best technical bid is awarded 100 per cent of the marking criteria and the prices are then adjusted proportionately (see end of Additional Data 3).

h)   The Most Economically Advantageous Tender or the lowest evaluated tender, after satisfying its technical acceptability. This involves evaluating every aspect, for example, schedule, management commitment, project personnel, design capability and procedures, and so on. In principle, it is similar to a point scoring system but endeavours to eliminate some of the subjectivity in awarding points by allocating a financial value to each element of the tender. A differential cost figure (either positive or negative) is allocated to each of the chosen aspects relative to the client's historical norms (performance above average, average and below average). The evaluations are added up and the adjusted lowest evaluated tender is chosen. This method, combining both performance and cost to the client, is to be recommended, and is discussed in more detail below.

i)   The lowest evaluated tender based on life cycle costings. This method adjusts the tender prices as a result of operating and maintenance costs. Whilst this approach takes a more correct long-term view, my experience is that owners have talked about life cycle costing for over 30 years, but have rarely used it. The main reason for this is because the client team (representing the owner) is more focused and rewarded on the short-term cost and schedule targets.

EVALUATING TENDERS

A further John Ruskin quotation:

It's unwise to pay too much, but it's worse to pay too little. When you pay too much, you lose a little money – that's all. When you pay too little, you sometimes lose everything, because the thing you bought was incapable of doing the thing it was bought to do. The common law of business balance prohibits paying a little and getting a lot – it can't be done. If you deal with the lowest bidder it is well to add something for the risk you run, and if you do that you will have enough to pay for something better.

As indicated in this quotation the evaluation process should, in effect, be a risk analysis of the tenders. *The results of the evaluation need to provide one with the greatest confidence that one has chosen the lowest risk combination of technical and project execution proposals, and contractual and commercial terms.*

In evaluating the tenders one seeks to eliminate the risks that have been identified. The evaluation process thus becomes a negotiation with the tenderer to provide alternative technical solutions or execution methods to remove these risks. The objective of post tender negotiation is to upgrade proposals where they are deemed to be less than optimal. It is not about reducing the tenderers' price, profit or fee.

The emphasis in the evaluation process will vary with the type of material, equipment and contract for which the tender or proposal has been requested.

Where the client is taking the risk and the supplier or contractor is reimbursed their costs, the supplier or contractor is, as has already been explained, supplying a service. In these circumstances the evaluation will be focused on the quality of the personnel and the working methods.

Where the client has passed control of the risks to a supplier or contractor, the supplier or contractor is responsible for delivering a finished end product. Here the evaluation will be focused on confidence in delivering in accordance with the schedule, and price

Consequently, the assessment of the two basic forms of contract is significantly different and must focus on different issues. However, in the 2000s there will always be an emphasis on health and safety, security and the environment.

Fixed price quotations for materials are, understandably, the most straightforward to evaluate. Check that the materials conform to the right specification and standards. Verify that the correct certificates have been provided. Finally, ensure the supplier has agreed to the client's terms (see Battle of the Forms, Chapter 6). Then, provided that the delivery is within the requirements of the project schedule, one can probably take the lowest price (within budget).

Material Requisitions are prepared by the specialist design groups and are usually requisitioned by commodity in categories or classes. For example: carbon steel pipes, carbon steel flanges, carbon steel fittings; alloy pipes, fittings and so on. These are then consolidated by buyers into a Request For Quotation, and issued to fabrication mills and stockist suppliers. The tenderers will be requested to provide individual prices as well as an overall total price. After the technical evaluation the commercial evaluation will probably just look at a comparison of the totals in order to identify the lowest quotation. However, the lowest price can be reduced if consideration is given to the best combination of commodity group prices and even selecting individual items within the groups. This is often termed 'cherry picking'. The result will be that the lowest price for the whole scope is achieved by placing two or even three (never more) purchase orders. This approach can also provide significant time and money benefits later when the dreaded 'new items' start to appear from the design groups. If only one purchase order has been placed then the new items have to go through the whole tendering process again. Whereas with three suppliers, the additional prices can be requested by e-mail or telephone and the quotations couriered, concluding the whole process within a week.

Equipment should have been specified using a functional or performance specification. Whilst this makes the specification process more appropriate and easier, the evaluation process requires more effort. Consequently, the focus of the evaluation effort will be on the technical features of competing designs. Checks also need to be carried out to ensure that the equipment will perform economically at the desired operating parameters.

Similarly, where a contractor is managing the project risks for a fixed price then, the emphasis of the review is on compliance with the scope and specification and review of the schedule. The client must request a contractor to cost and identify schedule effects for any differences or omissions with the project scope or specification. However, the most significant effect on a project's outcome is the contractor's execution capability. Consequently, further evaluation will seek to upgrade the contractor's:

- execution plans or method statements;
- schedules;
- organization of engineering and construction supervision;
- quality of key project personnel;
- availability of manpower.

Differences between the tenders need to be identified and evaluated. For example, the client needs to judge the value of the differences in the project schedules. A more difficult evaluation is the differing capabilities of the contractors, which will influence the quality of the product and the reliability of the stated schedule. Any contract exceptions need to be analyzed and recommendations made to the owner where the contractor will not accept the requested terms.

For a duty specification one must also be satisfied with the process, performance characteristics and economics of operation.

Provided that the product meets the specification requirements and has the best or acceptable delivery schedule, the lowest price becomes the key decision criterion.

Where the client is managing the risks and costs of a reimbursable service then, the emphasis of the review is on the strengths and weaknesses of the people and systems of the organization. The objective is to analyze the contractors' systems and execution proposals to quantify their effect on the project cost and schedule. Whilst the client can influence execution plans they will not be able (within the life of one project) to make significant changes to the effectiveness of the systems.

It should be remembered that the services required are technical and project management. How the contractor manages their administrative functions is up to them. Further, in the electronic age of the 2000s, reproduction and computer charges should never be on a rates basis but should be treated in a similar manner to other administrative costs. Consequently, administrative and computer costs should be included in a fixed fee.

Much of the following few paragraphs is based on, not only being debriefed by Exxon, but also on notes taken of presentations made by senior contracts managers on how they evaluate and select main contractors.

The total value of a reimbursable cost contract is made up of the following three elements:

1. materials and subcontracts;
2. services reimbursed at cost;
3. the contractor's overhead and profit, and other elements, or fee.

Items 1) and 2) are influenced by the contractor's capability to perform and items 2) and 3) are influenced by the contractor's quoted commercial terms.

The evaluation is based on what is termed 'hard money' and 'soft money'. The soft money is that which is affected by a contractor's performance. The hard money is a strict like for like comparison of the forecast contract value using the client's figures for those elements of the project that are based on estimates. For example, the forecast contract value is the sum of:

> a contractor's fixed fee including: administration, reproduction and computer costs:
> + hourly rates x client's estimate of reimbursable cost man hours;
> + client's estimate of the cost of materials and subcontracts;
> + an estimate of the cost of the impact of a contractor's exceptions to the contract terms.

The soft money is determined by evaluating the tenderer's *relative* ability to perform various project activities:

- detailed design;
- procurement;
- project controls;
- construction management.

The quality of the project manager, project team and other project specialists is also evaluated. The evaluations seek to determine in depth capability; for example, during the project launch phase, the execution phase and during the project close down phase. Finishing projects is much more difficult than starting projects. Consequently, how many projects has the project manager finished?

The ratings (exceptional, above average, high average, average, low average, below average and unacceptable – the normal range being below average to above average), for the activities identified above, are translated into effects on the following specific areas of project cost (either positive or negative adjustments or differentials – not total costs):

a. design manhours;
b. bulk materials;
c. equipment;
d. subcontracts;
e. construction supervision;
f. construction overheads.

- The cost of the design manhours, item a, is influenced by a contractor's productivity in the design function.
- The cost of bulk materials, item b, is influenced by a contractor's design capability and their purchasing and cost control capability.
- The cost of equipment, item c, is influenced by a contractors design capability and their procurement capability.
- The cost of subcontracts, item d, is influenced by a contractor's design, procurement, cost control and schedule control capabilities.
- The cost of construction supervision and construction overheads, items e and f, are influenced by a contractor's design, cost control and schedule control capabilities.

Thus design capability is crucial since it influences the cost of all items b, c, d, e, and f. It must be remembered, though, that this quantification is dependent on good client historical data of contractors' performance.

The positive and negative cost adjustments are summarized to provide the evaluated cost of the contractor's performance. To this is added the client's valuation of schedule differences. The lowest evaluated or conditioned tender is the sum of the hard money and soft money.

It is difficult for any purchaser to satisfactorily evaluate the complex mix of rates involved in projects, unless a complete scope breakdown is available. If a complete scope is available then a fixed price can be requested. When work cannot be quantified due to the early state of a design there may be no alternative to a Rates Based Contract. In discussing the dilemma of evaluating competing tenders with similar rates with my project team, I remember asking what would make one contractor more successful than another. We concluded that the construction manager and planner were the key people. We interviewed them, made our selection and wrote the names of our choice into the contract (with significant penalties if they were removed without our agreement).

An alternative approach to selecting contractors for project execution, used by BP, has much to commend it. Having obtained project approval there is some basic design work to be carried out in order to investigate alternative approaches for the development of the project. Accordingly, the client selects two contractors and provides each of them with a sum of money to carry out studies and develop their own particular ideas in a defined timescale. A small client team is allocated to work with each of the contractors. An evaluation is then carried out to determine which contractor has produced the 'best and most economic' solution. However, and most importantly, the contractors are evaluated on the basis of their behaviour. A selection is made on the basis of which one the client can work with, most effectively, during the course of project development.

DETAILED APPRAISAL ISSUES

The detailed issues that need to be examined have been tabulated below in generic lists, under the main appraisal categories. They attempt to cover both goods and services.

Technical and execution appraisal

- General presentation: the tender should be equivalent to the supplier's or contractor's best work. Thus, the quality of the document submitted gives a good overall idea of the quality that they will produce for the proposed product or service.
- Completeness: the tender may have been submitted stating that it was complete, but is it? Adjustments may have to be made to enable a like for like price comparison.
- Exceptions to the scope of work or specifications: there had better not be too many of these, and the reasons why need to be investigated. Again adjustments will need to be made to the pricing for a like for like comparison.
- Alternative proposals: the balance between efficiency, operability and reliability will need to be analyzed. Evaluate the technical and operational advantages and disadvantages of novel features. Check user references for where they have been used before.
- Safety and Quality Plans: these are crucial to the long-term operation and maintenance of the plant. Further, the supplier's or contractor's safety plans will impact on productivity and their ability to complete the facility on time.

    Has a higher than necessary quality been offered and, consequently, could the price be reduced for the right quality?
- Operability: the equipment needs to be analyzed to check that it can meet the performance specification criteria. Is the equipment operating at the limit of its capability? Are the efficiencies satisfactory? What guarantees are offered? Check any calculations and review any allowances or safety factors. Has the plant or facility been designed with operations and maintenance requirements in mind, or has it been developed just to suit a speedy construction schedule?
- Proposed programme: check the key dates. Has the programme been developed using a critical path network diagram? Is the logic used on the critical path valid? Has the delivery or completion date been achieved using impractical durations? Are there material constraints? What are the long lead items and are procurement cycles realistic? Check the details of critical design and construction activities.
- Subcontracting: it is accepted that tenderers might need to seek help for specific areas of specialist expertise to enhance their capability. However, too much subcontracting negates the whole process of pre-qualification.

On the other hand, have a sufficient number of local contractors been included in the proposal? Firstly, in order to reduce the amount of hard currency required and secondly, to act as a process of technology transfer. Giving local contractors experience of working with a more experienced contractor will enhance the country's skill base.

- Workload: as a client one requires a somewhat unreasonable situation. I should like the supplier to be operating at less than optimum capacity and yet remain a viable organization. I do not want the proposed order to form a disproportionate part of the supplier's workload. Similarly with a contract for services. The contractor's workload should be at, say, 60 per cent. Our contract will take up the next 30 per cent and we want the contractor to leave the remaining 10 per cent spare. Otherwise how will the contractor's systems cope when they are at full stretch and there is some extra work to be performed? Where will the extra resources come from?

  A contractor's resource numbers will need careful and detailed examination. They will spend considerable effort to demonstrate how their current workload is dropping off and that the proposed project will just fit appropriately to provide a consistent flat resource profile.

- Construction Design and Management Regulations, (also known as CDM): who is to be appointed to, what is termed, the construction design and management coordinator role (the old planning supervisor role)? Does the contractor have a sufficiently resourced team to carry out the project? Are they competent to do the job?

- Key staff: review the curriculum vitae of the project manager and their key team members. Check out their references – on one occasion a client spent a week telephoning Turkey in order to check my client reference. I was impressed with their seriousness. There may also be other personnel whose experience needs checking. This might be a technical specialist or other individual whose function is crucial owing to the nature of the project.

- Agency Staff: in the 1980s if the contractor had more than 5 to 10 per cent agency staff they would not be awarded the contract. In the 2000s, agency staff can be as high as 60–80 per cent. Unless these self-employed agency staff have been with the contractor for many years, they cannot be as effective as the contractor's own staff. I am a great believer in low agency staff ratios.

- Training: in less developed countries, training of the owner's staff will be a major consideration in the evaluation. Inspection activities can also form part of the training of a client's personnel.

- Evaluating the culture of the project team: choose people who have worked in the appropriate culture before. If the client takes the risk with the contractor providing a service, a cooperative culture can be developed. On the other hand, if the contractor manages the risk and is responsible for delivering an end product, a confrontational culture may develop.

  The appropriate culture is particularly important in a Joint Venture (also known as a JV). Most joint ventures are put together specifically for a particular project and are composed of very different contracting disciplines. It is, therefore, essential to evaluate the different entities' ability to work together. Once again the Scottish Parliament architectural joint venture (Scottish-Spanish) is an example where it did not really work.

- Performance tests: for operating or overload characteristics, exhaust gases, effluents, noise levels or vibrations. Agree the method of testing and any measurement allowances as early as possible. Although this is more important for the supplier and contractor, rather than for the owner.

- Vendor serviceman: is a vendor serviceman required for installation as well as commissioning, and have separate rates been quoted? Check the call-out notice requirements.

- Special tools: are any special tools required for operation and maintenance, apart from those provided by the vendor serviceman during installation?
- Testing and inspection: it is best done by the 'users', which for an owner client will be their operations and maintenance personnel. For a contractor, acting as a client, the users will be the installation and construction site engineers.
- Documentation: as a generalization, designers ask for too many different drawings. Do not ask for more information than you can analyze in the time you have allowed, for performing the tender analysis and review. Make sure that a meaningful payment is tied to the provision of important documentation.
- Certificates: contractors purchasing on behalf of Middle Eastern clients may have to obtain certificates of origin. Inspection and test certificates will be needed for maintenance manuals, so check that a sufficient number of copies are provided.

    Are certificates for testing and lifting equipment and pressure vessels, and so on. available?
- Painting, packing and weather protection: check for any specialist treatments. Check for any constraints for storage on site or opening of crates.
- Spares: for the construction, commissioning and 2-years operation will have been specified, and should have been normalized is some way. Check for any that are particularly long delivery items.

    However, the evaluation needs to be done separately. If spares are not evaluated rigorously they can (and can be used to) distort the overall evaluation process. As one of my colleagues[4] said to me, 'A qualified pump manufacturer does not knowingly supply a pump that uses twice as many bearings as their competitor's. On the other hand, a less scrupulous supplier may well put forward a more "optimistic" list of very high priced spares.'

Contractual appraisal

- Exceptions: the terms submitted as part of the enquiry are the basis on which you, the client, want to do business. As one client said during a debriefing, after losing the second tender in a row, 'If you want to do business with us you had better learn to accept more of our terms and conditions.'

    Does the tender letter sent with the offer qualify it, or modify the offer, in any way?
- Documents signed: in the haste to submit a tender this aspect can get overlooked. For the tender to be a valid offer it needs to be signed by an authorized signatory or officer of the company.
- Secrecy and confidentiality: is the secrecy or confidentiality agreement signed as accepted, or are there any qualifications?
- Validity: the validity period should be long enough to enable the client to carry out the evaluation, chase up queries, attend presentations (see below) and negotiate with at least two tenderers. Further, the client has an obligation to complete these activities in the requested validity period and not to overrun and, thus, request the tenderers to extend the validity of their proposals.
- Key dates: the dates should be consistent throughout all of the documentation.
- Payment terms: check that they have been provided in the requested format and check the arithmetic.

---

4   Vernon T Evenson, Project Manager.

- Cost Price Adjustment (also known as CPA): if the economic climate justifies the transfer of this risk to the client, what formula and base date has been suggested and what indices are proposed?
- Incentives and Liquidated Damages (often referred to as LDs): provided some benefit can be identified, consider offering an incentive mechanism for early delivery rather than the negative attitude of damages for late delivery. Remember that liquidated damages can be used for the delivery of any part of the order. For example, the provision of documentation or the date for an inspection or test visit. Cancelled test visits involve the client in significant costs. Has the tenderer put a limit on the liquidated damages?

  Use the liquidated damages to test the supplier's confidence in meeting the schedule and completing on time. Run through the following role play during the final negotiations, 'The good news is that you are one of the last two tenderers being considered. The bad news is that liquidated damages have increased to twice the amount' and watch the manufacturing or operations directors' reactions. If the proposed supplier or contractor start to protest or argue, the implication is that they think they might have to pay them – they think they might overrun the schedule. The reaction that one is looking for is the supplier who says, 'That's alright because we will be on schedule and will not be paying any liquidated damages.'
- Subcontractor Terms: are the terms and contractual arrangements with the subcontractors compatible to the project objectives?
- Insurance: so as to avoid delays to the project, check that lost and damaged items will be repurchased regardless of any insurance claim. Obtain a copy of the certificates for required project insurances.
- Warranty: what are the terms and what is the period of the warranty? If replacement parts are required for equipment the warranty period should be extended, at the least, for the replaced part. Similarly, make sure that the warranty period or date is extended for every extension to the project schedule as a result of extra work.
- Bonds and guarantees: check that the requested documents have been provided and get your specialist to check the wording.

Financial appraisal

- Price: is the price in the requested format and is it arithmetically correct? What has been omitted in order to make the price appear competitive? For example, are capital spares included? How do any proposed alternatives effect the price?

  Are prices for changes, extras or spares disproportionately high?

  Is the price affected by local income tax or other taxes?
- Delivery and Incoterms: does the price include delivery in accordance with the requested Incoterms? Are there any additional import duties, taxes or other customs clearance costs?
- Fees for Reimbursable Cost Contracts: these should be all inclusive and fixed. Review and discuss the make up of the rates so as to be satisfied that the rates represent the net cost to the tenderer. If any profit elements are discovered hidden in other rates, the contractor should be disqualified and crossed off any future tender lists.
- Overheads compared to reimbursable costs: review and examine the definitions for personnel and for services. Make sure that overhead costs are not being camouflaged as reimbursable costs.

- Overhead and Profit (also known as OH&P): check that there is no overhead and profit being charged on overtime. Overhead and profit should be recovered through the agreed normal working hours.
- Exceptions: are any aspects of the specifications regarded as extras to the price?
- Quality Assurance (known as QA) and Expediting: include *estimates* for quality audits and expediting and inspection costs.
- Environmental: are there any costs associated with packaging disposal and recycling?
- Value Added Tax: in the European Union, value added tax paid by a business is offset by value added tax recovered when selling goods or services to customers. Consequently, value added tax is not an issue. However, what are the arrangements when dealing with foreign countries outside one's home country environment?
- Exchange Rate: what is the currency for payment of the price? Project exchange rates should be defined and agreed so that comparisons can be made with those quoted. Can purchases made in a foreign environment be made in the local currency in order to reduce the amount of hard currency required?
- Payment Terms: in order for the payment terms to be compared on a like for like basis, they will need to be discounted to the contract base date using Discounted Cash Flow (also known as DCF) or Net Present Value (known as NPV) techniques.
- Operating Costs: cheap equipment, which is not very efficient, needs to be compared with more expensive equipment that will cost less to run. This comparison should also be carried out using net present value techniques, rather than a simple extension of extra fuel costs over an arbitrary number of years of operation. Having said this, when evaluated properly, the differences tend to be small.
- Bonds and guarantees: there may or may not be a fee included for a parent company guarantee. On the Scottish Parliament project only one tenderer included a fee (1 per cent of the anticipated construction cost of £50 million) the remaining tenderers providing it at no charge. There will, however, be costs associated with bonds provided by banks and these costs will depend upon market rates.
- Inflation or cost price adjustment: calculate the effect of inflation over the duration of the contract period and adjust the price accordingly.
- Financial Status: a credit rating agency's report should be a routine check. However, it is of questionable value since a company can still go bankrupt despite a positive report. Determining the ultimate holding company is of paramount importance if guarantees are to have any substance.
- Subcontractors: has the tender included data on the financial status of proposed subcontractors?

Summarizing the results

The results of the analyses are then summarized on a Tender Evaluation Form, or as it is usually termed, a Bid Tabulation. A proprietary example, for materials and equipment, is shown in Figure 11.1. Naturally, this document is in an electronic format with all the functions and features of a spreadsheet.

Page 1 lists all of the items requisitioned with reference numbers, quantities, unit prices and so on. Page 2 adds other items that will be included in the final purchase order price. For example, inspection, packaging, spares and special tools. The right hand side of pages 1 and 2 has space for comments, remarks or notes.

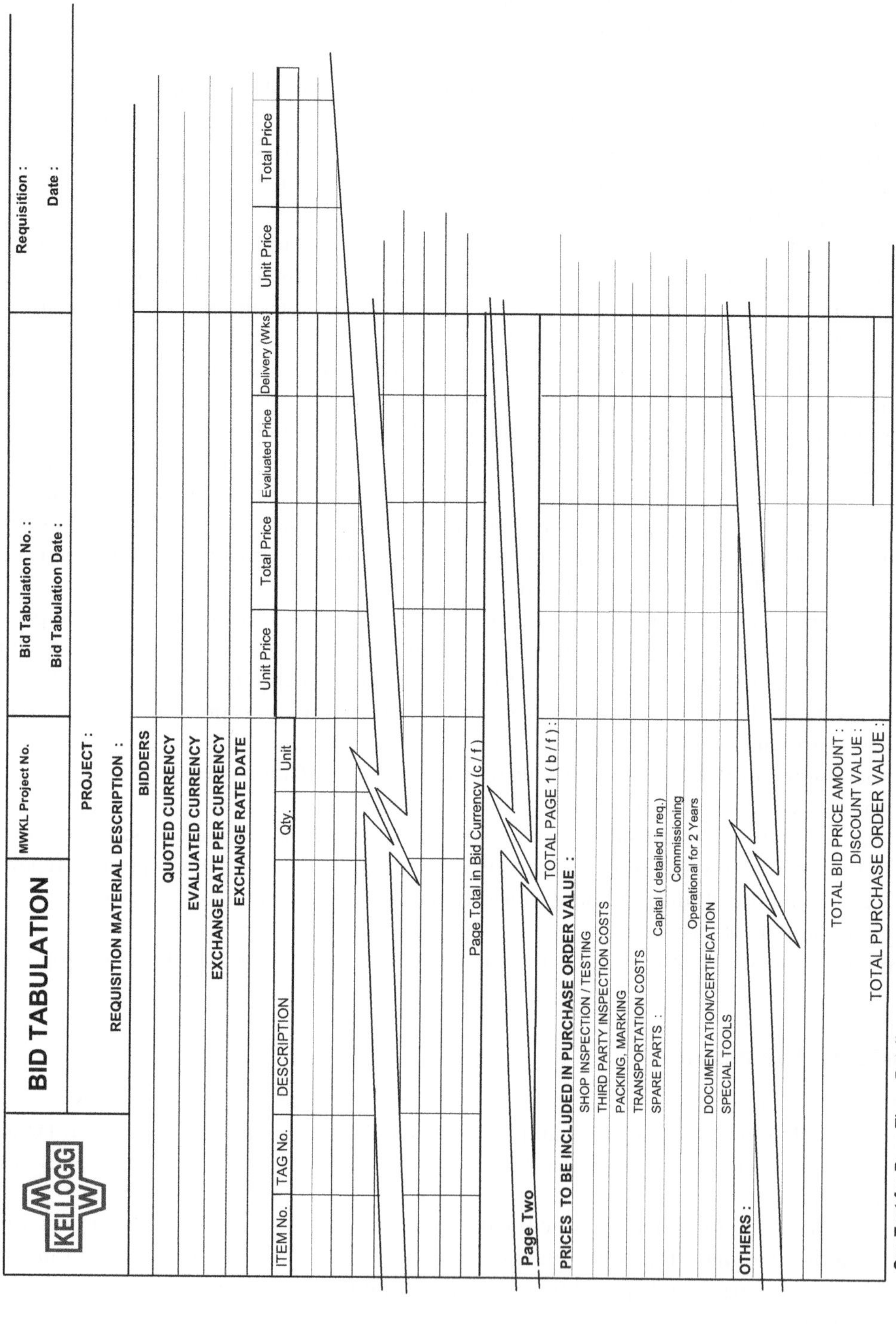

See Text for Page Three Details

Figure 11.1 Tender evaluation form

Page 3 then analyses all the other issues that effect the evaluation and eventual selection of the successful tender. The main headings on this page are:

1. Details of manufacture, delivery and duties.
2. Estimated shipping weight, dimensions and volume.
3. Date for key drawings.
4. Acceptance of key purchase terms. For example, packing instructions and services and conditions available at the project location.
5. Payment terms and evaluated cost of progress payments.
6. Warranty and percentage of total order value. Length of warranty period.
7. Liquidated damages for late delivery or other key event. Maximum limit of damages.
8. Tender reference, date and validity expiry date.
9. Confirmation of technical acceptance.
10. Health, safety and environmental comments.
11. Total evaluated price.
12. Finally, the recommended supplier and the reasons for the selection.

On page 3 the right hand side is used for the following:

- Required on site date.
- Budget.
- Selected supplier
- Tenderers not included in the evaluation and the reasons.
- Check boxes for the following capability validations:
    - financial situation;
    - workload;
    - references;
    - supplier information system.
- Buyer signature and date.
- Project equipment engineer and procurement manager: Name, signature and date.
- Manager of equipment engineering and procurement: Name.
- Project manager: Name, signature and date.

In the case of a contractor acting on behalf of an owner there will also be space for their name, signature and date.

Presentations and interviews

The evaluation process is not complete until the client's selection is confirmed by a visit to the supplier's or contractor's premises.

For key equipment a visit to a supplier might be set up at a working level on the lines of, 'Our project manager would like a meeting to finalize the details of the order and see your facilities.' If the managing director and manufacturing director do not attend such a meeting, then it is a clear signal that the business is not sufficiently important to them. It is at this meeting that the scenario covering liquidated damages (described under Contractual Appraisal above) should be played out.

A visit to a contractor's offices enables the client to see the contractor's facilities first hand and the support that the organization provides to the project team. The agenda for the visit, focused primarily around a presentation covering the contractor's execution plans, is advised in advance. Who should make the presentation, and who should attend on behalf of the contractor, is not spelt out. This enables the client to evaluate the contractor's attitude to the proposed project. Is the presentation carried out by line or functional managers, by a group of people selected for their presentation abilities or is it performed by the proposed project team? Further, if the presentation is performed by the project team, do the functional managers and, more importantly, the chief executive appear at some stage to demonstrate their commitment to the project? Upon arrival the client team should make changes to the agenda. This enables the client to evaluate how the contractor copes with the unexpected. Interviews should also be carried out, not only with the key team members, but also with lower level specialists and designers. This enables the client to determine the experience and commitment in depth within the organization.

REACHING A CONCLUSION

Because quite different aspects need to be compared, actually reaching a conclusion on the preferred tender is difficult. Balancing a contractor's greater project management capability against another's superior technical solution is not easy. This dilemma can be resolved by impressions made in the presentation. Is the proposed project team genuinely committed and enthusiastic? The team that is truly hungry for the project – that really wants the work, will be the most motivated to complete the project ahead of time. Finally, as indicated in the Additional Data that follows, choose the better project manager.

The Additional Data tabulations 1, 2 and 3, show that the evaluation criteria are of such complexity, that the results produced from the elaborate computations can be of dubious validity. Dr. Roy Whittaker's 5 view was to, 'Try as many different schemes as possible and hope that all of them – including a purely subjective appraisal – give the same result.' Surveys showing overall winning factors are tabulated in Additional Data 3.

Finally, unsuccessful tenderers should be advised and offered a debriefing by the client. The benefit to the client is that the supplier or contractor will provide a more effective or competitive tender for the next enquiry.

ADDITIONAL DATA 1

Modified extract from a paper on 'Contract Award Practices'[6] in civil engineering:

- United Kingdom   The lowest bid provided it is not less than 90 per cent of the average of all bids.
- Wales   The second lowest tender.
- Denmark   The tender nearest the average after rejection of the two highest and two lowest tenders.
- Germany   Selection from the proposals by subjective preference of the client not by objective decision rule.
- Italy   The tender closest to the average of all the tenders received.
- Japan   The first stage of work is split between three contractors. All of the contractors must tender for the entire job when tendering for the first stage. The best performer then wins the rest of the job.
- Okinawa   The tender nearest to, but below, the average of all tenders.

5   Project Group Manager of the old ICI
6   By T.M. Lewis University of the West Indies, Trinidad. Undated but with a reference dated 1979. The above extract is referenced H.W. Hunt et al, 'Contract Award Practices', *Journal of the Construction Division*, ASCE, Vol. 92, No. C01, January 1966, pp. 1–16.

- Korea   The tender nearest the average after rejection of the highest and lowest tenders.
- Pakistan   The lowest tender provided it is not less than 80 per cent of the client's estimate. Otherwise the next lowest (with the same proviso).
- Philippines   The tender nearest to, but higher than the average, provided it is less than the estimate. Otherwise the tender nearest to but below the estimate.

Another old paper[7] sets up a matrix with two sets of variables. 'These are: (a) qualifications based on information in the tender, and (b) qualifications based on price. A score of 0–70 is awarded for either type of qualification. These scores are plotted on a (diamond shaped) matrix. The higher up a tender is placed the greater the prospects for a successful contract.' The (a) factors are:

1 Project Understanding	2 Approach	3 Compliance with Specs	4 Specific Experience	4 Management Qualification
20	10	10	15	10
U P Av G E	U P Av G E	U P Av G E	U P Av G E	U P Av G E
2 5 10 15 20	1 3 5 8 10	1 3 5 8 10	1 4 7 10 15	1 3 5 8 10
U: Unsatisfactory, P: Poor, Av: Average, G: Good, E: Excellent.				

ADDITIONAL DATA 2

The following is the Criteria and Weights for Major [Evaluation] Factors from a case entitled Teem Aerospace Corporation.[8] This case provides a thorough examination of the strengths and limitations of a point scoring system for tender evaluations.'

Technical	**Management**	**Financial**	**Manufacturing**	**Quality Control**
40 Design approach	20 Plan	30 Price	20 Departmental experience	20 Policy
25 Technical capability	26 Organization	15 Strength	15 Plan	25 Operational system
15 Development plan	14 Manpower	15 Accounting system	13 Facilities	15 Technical capability
8 Reliability	20 Controls	20 Cost control	6 Skills	10 Facilities

---

7   'Developing a Matrix for Tender Evaluation' by G. Peters of British Gas Corporation, pp. 1003–1005. Undated and source unknown.
8   'Cases in Manufacturing Management'. By Albert N. Schrieber. The McGraw-Hill Book Co, 1965. pp. 385–386. Library of Congress catalog card number 64-8623

7 Specification conformity	5 Experience	10 Estimating technique	10 Tooling	15 Reliability
5 Release of proprietary rights	10 Reliability		16 Controls	
	5 Labour relations		7 Improvement	
			5 Training	
100 Total	100	100	100	100

Each of the main criteria are split into subcriteria. The management subcriteria are identified as follows:

			Points
I.		Management plan	20
	A.	Adequacy of plan – consistent with program requirements; depicts understanding of problems with logical solutions and actions.	10
	B.	Compliance with proposal instructions.	3
	C.	'Make or buy' plan (including procurement).	7
II.		Management organization	26
	A.	Corporate structure.	5
	B.	Specific management organization, depth, lines of authority, experience.	12
	C.	Organizational stability.	4
	D.	Support organization.	5
III.		Manpower	14
	A.	Requirements and availability.	7
	B.	Manpower acquisition plan.	5
	C.	Effect of current and future business on manpower program.	2
IV.		Management controls	20
	A.	Adequacy of management controls (time and dollars): type, frequency, effectiveness.	12
	B.	Top-management participation.	5
	C.	Corrective action by management: what, how, effectiveness.	3
V.		Management experience	5
		Similar projects or jobs successfully managed of a similar complexity and magnitude.	5
VI.		Reliability	10
		Management participation:	
	A.	Policy.	3
	B.	Procedures.	3
	C.	Organization.	4
VII.		Labour relations	5
			Total 100

Some more recent and actual weightings, based on Exxon and public sector experience,[9] are listed below:

Post Tender Interview Weighting Factors:
FIXED PRICE:

Technical ability		25%
Understanding	5%	
Method statement	10%	
Special skills	5%	
Procedures	5%	
Team		75%
Project manager	50%	
Team	25%	

REIMBURSABLE:

Technical ability		55%
Understanding	10%	
Method statement	10%	
Special skills	10%	
Procedures	25%	
Team		45%
Project manager	25%	
Team	20%	

The reason for the difference between the percentage allocations for the two contract types is the nature of how control is exercised. For a fixed price contract the client is totally dependant on the contractor's team managing the project according to their method statement or execution plan. For a reimbursable contract the client team exercises more management control and is dependant on the effectiveness of the contractor's systems and procedures.

Any weighting system needs to be adjusted to suit the particular circumstances of the enquiry process. For example, 10 per cent could be allocated for a presentation or 10 per cent allocated for relationships, attitude and dedication – the remaining percentages would then be redistributed accordingly.

An Esso internationally advertised pre-qualification[10] stated that, 'Bidders will be evaluated on their perceived willingness, documented track record and demonstrated plans to fully comply with the [Government] Directives and [local] company requirements.' These directives required, 'that all contractors create value in country with targets of 45% by end of 2006 and 70% by end 2010 in terms of total monetary expenditures through the deliberate utilization of [in country] human and material resources without sacrificing safety, health and environmental standards.'

The UN International Trade Center (ITC) awards contracts on the following basis:

- RFQ[11]: 'Awarded to the lowest technically acceptable offer.'
- RFP: 'Awarded to qualified bidder whose bid substantially conforms to the requirements set forth and is evaluated to be the lowest cost.'
- ITB: 'Normally price is the sole determinant in making an award. Where all technical criteria are met, award is made to the lowest bidder.'

---

9   Presented as part of a lecture (in the 90s and 20s) at Cranfield School of Management on 'How Clients Choose Contractors', by Nigel Parry. One time Project Manager with Exxon and independent Consultant in the public sector.
10   *Financial Times*, June, 2006.
11   See Chapter 9 for the ITC definition of these terms.

ADDITIONAL DATA 3

A 'Survey of Engineers, Managers Rates European E&C[12] Companies.'[13]

During the summer of 1993...a management consulting firm...interviewed 62 Western European downstream 'owner decision makers'.

Study results indicate: the factors that influence contractor choice...for refining and petrochemical projects. ...The primary areas of responsibility of those that were interviewed for the study are:

- *Project management*   *34%*
- *Engineering*   *27%*
- *Construction*   *19%*
- *Procurement*   *9%*
- *Senior management*   *8%*
- *Planning*   *3%*

The interviews were conducted in Belgium, France, Germany, Italy, the Netherlands, Norway and the UK.

The table lists the most important factors influencing the choice of an E&C company and shows how this list of factors has changed since 1980.

COMPARISON OF BUYING FACTORS – 1980–1993

Buying factor	1993 rank	1993 avg. rating	1990 rank	1990 avg. rating	1980 rank	1980 avg. rating
Key personnel assigned to project	1	9.36	2	9.58	1	9.34
Project management capability	2	8.79	1	9.65	2	8.48
Project control systems	3	8.26	5	8.65	5	8.10
Evaluated price of E&C services	4	7.84	7	8.10	4	8.10
Responsiveness and flexibility	5	7.79	3	9.10	9	7.66
Detailed engineering capability	6	7.74	4	8.77	3	8.34
Quality of proposal	7	7.66	12	7.10	N/A	N/A
Experience with similar work	8	7.57	6	8.5	5	8.00
Quality of senior management	9	7.45	11	7.23	7	7.92
Size and location of office	10	7.02	13	7.0	N/A	N/A
Ability to do work in one office	11	7.00	9	7.54	13	6.82
Construction management capability	12	6.78	8	7.65	8	7.68
Procurement capability	13	6.66	10	7.29	11	7.04

---

12   E&C: Engineering and Construction.
13   *Oil & Gas Journal*, Jan. 10, 1994

Concep./front-end engineering capability	14	6.60	18	4.54	15	6.34
Experience in same geographical area	15	6.41	16	6.54	10	7.06
Contractor's total man hour estimate	16	6.26	15	6.70	N/A	N/A
Start-up/training capability	17	5.15	17	6.18	16	5.63
Capability of sales representative	18	4.73	14	6.93	17	4.80

In 1988 one of my MSc in Project Management students, Michael T. Eaton, carried out a similar survey using semi-detailed questionnaires as part of his dissertation entitled, 'The Winning of Projects in the UK Engineering Construction Industry'. The survey endeavoured to determine why 16 UK (process industry) contractors were successful in being awarded 26 (competitively tendered) engineering, procurement and construction management projects, over a 5 to 6 year period.

The results are consistent with previous comparisons. However, some additional aspects were highlighted – specifically, the importance of the project manager. The overall average ratings are identified below with the frequency of the maximum rating of 4.0 (most important, crucial, primary competitive advantage) shown in brackets:

1. Overall execution plan and understanding of project scope/objectives.   3.52 (15)
2. Confidence, commitment and interest shown to client.   3.35 (12)
3. Experience and quality of nominated project manager.   3.27 (14)
4. Experience and quality of nominated lead/specialist engineers.   3.23 (11)
5. Experience and quality of nominated project management team.   3.19 (10)
6. Your rates, costs, charges.   3.16 (11)
7. Your fee.   3.14 (8)
8. Responsiveness, flexibility and thoroughness.   3.04 (7)
9. Project organization, definition of responsibilities and roles.   3.00 (8)

In 2001 an MBA student, Diana Bradshaw, carried out another survey as part of her project work, 'Winning International Aid Projects'. The project covered the winning of contracts (6 months to 3 years duration) for international aid related technical assistance for developing countries. The methodology and questionnaire were similar to Michael T. Eaton above but covered 12 organizations in the UK, Canada, Denmark, France, Germany and Italy. In addition, five banks and other donor agencies provided responses but were not included in the statistical analysis. The average ratings for bidding success factors were (unfortunately for a direct comparison with the Eaton results) based on a maximum score of 5 (very important). However, the similarity of the results is interesting.

1. Experience and quality of project manager.   4.8 (9)
2. Experience and quality of project team.   4.4 (6)
3. Experience and quality of project director. [A function of how consultants organize their work].   4.8 (9)
4. Overall execution plan and understanding of project objectives.   4.2 (5)
5. Project organization, definition of roles.   4.0 (4)
6. Confidence, commitment and interest shown to client.   3.8 (2)
7. Responsiveness, flexibility and thoroughness.   3.7 (1)

The price and schedule, together with project controls and reporting procedures, came equal eleventh in this survey.

Loan or donor agencies letting international consultancy work tend to declare the details of the evaluation criteria and point count system. Each technical criterion is allocated points out of 100 distributed between different subcriteria. The commercial comparison is then carried out and the lowest priced tender is awarded 100 points and the remaining prices are then adjusted proportionately. 'The technical and commercial evaluations are then combined on an 80/20 basis. 80 per cent of the technical proposal evaluation score is added to 20 per cent of the commercial proposal evaluation score. The contract is then awarded to the tender achieving the highest overall score.'

Paragraph 3.3.10.3 'Evaluation of offers', of the EC Practical Guide to contract procedures[14] sets out the details of how points are awarded and the calculations required together with examples.

---

14   See *Additional Sources and Contact Details*.

12 *Incentives*

The payment terms are extremely useful for the client as a negotiating device. However, as well as negotiating a better deal, the payment terms should be structured to operate as an incentive motivating mechanism. Firstly though, we need to examine how the payment terms can operate to the detriment of the client.

DISINCENTIVES

Well thought-out payment terms can make a significant difference to a contractor's gross margin and a company's cash flow. This can be achieved by allocating a disproportionate element of overhead, profit and costs to activities in the payment schedule at the front end of the project. Consequently, clients should be on their guard against a supplier or contractor front-end loading the progress payments.

The following is relevant to tender or proposal payment schedules and is primarily applicable to construction activities, although it can apply to the design phase or the supply of equipment. Plotting the work done (man-hours, money spent or percentage progress) against time gives an 'S' curve shape as illustrated in Figure 12.1. Line 1 represents the value of the work to be performed or the work scheduled. It represents what the client expects.

The actual payment tabulation that the contractor has proposed in their tender is shown as an escalator (lines 2a and 2b) with regular payment steps. Line 2a represents payments made on a monthly basis, and line 2b represents payments made on achieving a specific percentage progress. For simplicity the same degree of front-end loading has been used for both payment mechanisms. These regular payments have then been transposed into an S curve for ease of comparison with the work schedule curve. This front-end loaded curve is illustrated in Figure 12.1 as line 3.

For the purposes of my hypothesis a 15 per cent down payment for, say, site establishment has been used. Either set of payments, if written as figures in tabular form, might appear as reasonable. Drawn graphically the escalator of payments appears to show a balance between both contracting parties. The financing of the cost of the work, from one payment to the next, swings between client and contractor. When the step is horizontal the supplier or contractor is bearing the cost of the work, and when the step is vertical the client is making a payment. However, drawn as an S curve it is clear that the overall effect of the payments scheme is loaded in favour of the contractor.

Custom and practice in the construction industry is to measure the work achieved on a monthly basis. Thus, the supplier or contractor (despite the advance payment) could receive a payment in

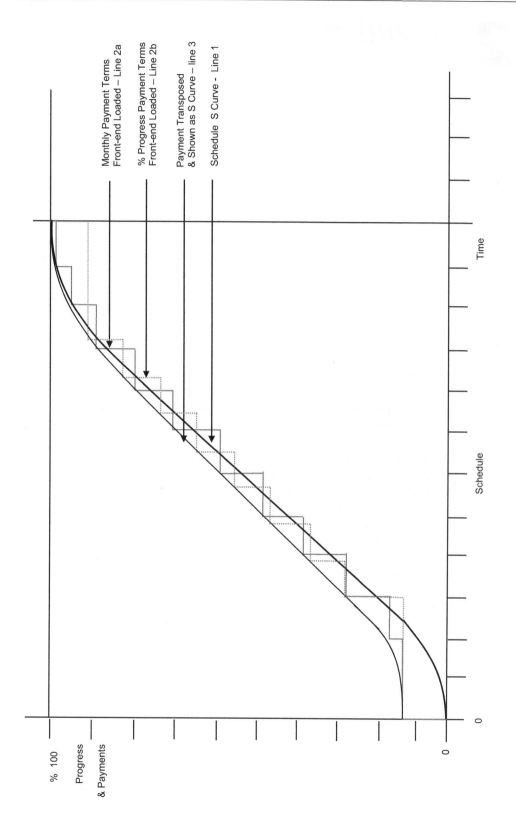

Monthly Payment Terms
Front-end Loaded – Line 2a

% Progress Payment Terms
Front-end Loaded – Line 2b

Payment Transposed
& Shown as S Curve – line 3

Schedule S Curve - Line 1

% 100

Progress
& Payments

Time

Schedule

0

0

Figure 12.1   Schedule and payment

the second month, and also continue to receive payments in the last two months of the project. I contend that these payments do little to motivate the contractor in these two crucial stages, namely the launch phase and close out phase of the project. Whereas, paying on achievement of targets, namely specific progress percentages, is motivating. Further, during the project launch and close out phases, there is a much clearer incentive to achieve progress at the start and to receive the final payment at the end.

Having set the scene, what happens in practice? My proposition is that advance payments and, or front-end loading of payments, will always make a project run late. Figure 12.2 shows the progress of the actual work performed (line 4) as a result of the front-end loaded payments (line 3). The rot sets in during the project launch phase and is unlikely ever to be recovered.

Having received an advance payment, I contend that the contractor's corporate and project management relax. The project is in the black and there is no need to be concerned about it. Leave the project to get on with things. The result is that progress in the launch phase is less than optimal. The end of the launch curve and the beginning of the straight line portion of the S curve is missed. Shown in Figure 12.2 as a shift from point x to point y – a delay of about 2 weeks. I will be over generous and assume that the project then takes off at the maximum rate as planned. On this basis the progress curve will proceed parallel to the schedule curve. In order to improve on this rate of progress, exceptional leadership activities would be required and this is unlikely in view of how the project started. Consequently, the project will run at least 2 weeks late. In addition, though, the front-end loading of the payments means that the actual money paid is more than the value of the work performed. So that, at about 60 per cent through the project, most of the overhead and profit has been extracted from it. As a result, the project receives less attention from management and the work force slows down. Further, by this stage the client is likely to have interfered in some manner (making changes or being late in providing information) and the workforce can see the end of the project and the end of their employment. In addition, the supplier or contractor might move resources on to other projects so that the whole process can be repeated. The result is that, with only a slight drop in the rate of progress, the project is likely to be 2 to 3 months late.

INCENTIVE MECHANISMS

Incentive strategies need to be determined as part of the contracting strategy and payment strategy discussed previously. Whilst an incentive strategy needs to be developed, it should be held in reserve until the supplier or contractor selection process has taken place and negotiations are well developed.

Organizational incentives

Unfortunately, there is no direct connection between a contract incentive mechanism and the transfer of this incentive to the motivation of the individual team members. Consequently, a company should be chosen whose personnel are already motivated to meet project objectives and a link needs to be created with the project management team and this can be achieved through two mechanisms.

One way to apply a project incentive scheme is to share a bonus amongst the project team through a formula or, alternatively, through performance appraisals of the individuals. From the contracting company's point of view this can be divisive. Why should one project have a bonus scheme and not another, in the same company? Of course, from a client's perspective, how a contractor runs a corporate bonus scheme is of little interest to them. A client wants to know that

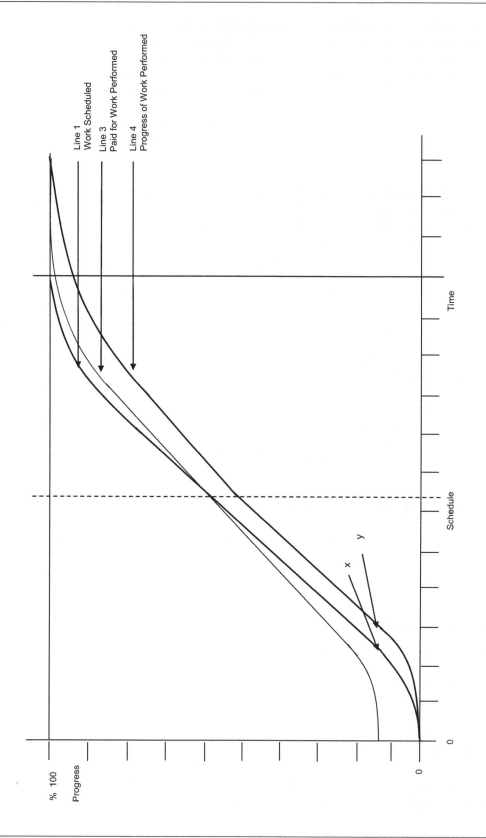

Line 1
Work Scheduled

Line 3
Paid for Work Performed

Line 4
Progress of Work Performed

% 100
Progress

Time

Schedule

x

y

0

0

Figure 12.2   Progress and payment

the people working on the project will feel motivated because they will be rewarded with some portion of the bonus if their project meets the performance targets. A solution is to have only part of the bonus payment awarded directly to the individuals on the project, with the major portion being awarded to the project, that is, the company. This has the effect of focusing management attention, support and encouragement for the project.

A company with an effective bonus scheme can develop a culture where the dominant performance criteria is meeting project objectives and having a happy client. If meaningful bonuses are then paid to individuals based on their performance, it can produce highly motivated individuals.

A corporate performance bonus scheme should not be based on a formula, whereby everyone shares in a corporate pot, since it eventually becomes an expected norm and its effect is dissipated. It should be entirely at the discretion of the top executive(s). It should be paid to those who perform even when their division's results are down as a whole but other divisions have generated the necessary profits for sharing out. This type of scheme and the commitment it generates is rare. Consequently, clients that institute project bonus schemes like to have a portion of any project bonus earned paid to individuals. However, with the type of scheme already described, clients should be willing to accept that a project bonus is paid to the company concerned rather than to individuals.

The second method of conveying the project incentive to the project team is through the project manager. Project managers, and the individuals reporting directly to them, need to be self-motivated to achieve targets and characters who will lead and inspire the rest of the people working on the project.

Further, in selecting a contractor, the project management team should be analysed to determine that it has been put together with people of the appropriate culture. It will probably be a mistake to select a team for a Fixed Price contract from people who have just finished a Reimbursable Contract, and vice versa. As already mentioned it takes the duration of one project to change peoples' behaviour from a reimbursable cost service culture to a fixed price product culture.

The conclusion is that a project incentive scheme must be based on teamwork, and the client must work with the contractor. This, in turn, implies either a Client Managed Risk form of contract or a Shared Risk form of contract. In working with the contractor or supplier the client must focus on restraining the contractor from driving too hard for the cost and schedule targets. The client, representing the user, must focus on contributing their expertise on operational issues – a skill and experience that can be missing in a contractor, however, less so in an equipment manufacturing organization.

Incentive techniques

Sports people and project people like to achieve and beat targets. So, the first mechanism to be used should be a characteristic inherent in the project itself – the use of milestones. Milestones are sometimes used in the belief that they are mechanisms for the control of progress. This is a mistake – if you are late arriving at a milestone, it is too late to do anything about it. Milestones are motivators and should be used as such.

Firstly, at a strategic level, use the Product Breakdown Structure (PBS) to create sub-projects whose completion can be used as targets.

Chapter 3 (Contract Strategies) used an example of two contractors performing similar work and being told that the best performing contractor would be awarded the subsequent remaining work as a follow-on contract or change order. This is easiest to apply in the building or construction industry where there are a number of similar units in a project. Nevertheless, it can also be applied to other project situations by breaking the project into its natural phases or stages. For example, a contractor might perform a feasibility study and, if the client finds that they can work effectively with them, the work for the subsequent phases can then be negotiated and converted into whatever form of contract is appropriate.

Secondly, at the tactical level, in negotiating the schedule (before contract award) make sure the supplier or contractor includes a number of milestones. The project team, having set these targets themselves, and having 'bought in' to the schedule, should be motivated to perform more effectively.

If you are able to influence the manner in which the project is controlled and administered. Persuade the supplier or contractor to use a 'war room' approach and display the project plans and progress charts. The programme should be displayed with the milestones highlighted and progress achieved indicated against the plan. 'S' curves showing percentage planned and percentage achieved together with productivity, for each discipline, can be very powerful motivators in this context.

When a supplier or contractor gives notice that they have achieved completion, and there is an expectation of a final payment and/or a bonus, there is an opportunity to focus their attention to a degree that may not have been possible before.

At the completion of any project there is always a list of small items that are not quite finished – a punch list. It is crucial that the client's team review this list with a fine-tooth comb. If the list is too long, or if there is any item of significance, then the claim for acceptance of completion should be rejected. This will usually result in major efforts being made to ensure that the next claim for acceptance is not rejected.

Finally, do not forget appropriate expressions of congratulations, at various intervals or key milestones in the project, to accompany any initiative. People like to be praised for a job well done. A letter from your managing director to a contractor's or supplier's managing director can have considerable impact, and reinvigorate their motivation. This should be used sparingly otherwise it becomes devalued.

The purpose of incentives is to move the project team's motivation up to, or to maintain their motivation at, level five of Maslow's[1] hierarchy of needs – Self-actualization (personal growth and fulfilment). The more this can be achieved by the incentive mechanism the more effective it will be. A project management team, in a company with a good project culture, should already be at level four – Esteem Needs (achievement, status, responsibility, reputation).

Contract incentives

The purpose of any incentive mechanism is to align the objectives of the contracting parties. However, traditional UK contracting culture has been to 'penalise' a supplier or contractor for being late through the use of Liquidated Damages. Rather than this negative attitude, a more positive approach is to reward the supplier or contractor when they deliver something, which is usable to the client, ahead of schedule.

An example of turning what are, in effect, damages, into an incentive, is the use of the concept of 'lane rental' for motorway maintenance. The longer the tenderer needs for the performance of the necessary work, the longer the client is deprived of its use, the more expensive the contractor's tender. Further, the more the contractor overruns their schedule for the work, the more damages they have to pay for the lane rental.

If how people are paid determines their behaviour, then the nature of the supplier managed risk (fixed price) contract should be self-motivating, if it is structured with 100 per cent payment upon completion. This is standard practice when purchasing materials. Of course, the ultimate of 100 per cent completion is any contract based on meeting performance related criteria, and this is normal practice for the purchase of equipment. Achievement of materials and equipment targets

---

1   A. Maslow's hierarchy of needs was proposed in a 1943 paper 'A Theory of Human Motivation'. His theory (depicted as a pyramid) stated that, 'as peoples' basic needs are met they need to satisfy higher needs.' Level 1 – Physiological (basic life needs – food, drink, shelter, warmth, sleep etc.). Level 2 – Safety (protection, security, law, order etc.). Level 3 – Social/Belonging (family, affection, relationships, work group etc.).

are relatively straightforward to measure and equipment guarantees should act as effective driving forces for the manufacturer.

However, unless the order is of major significance, it may not be a high priority within the supplier's overall portfolio of orders. Consequently, an additional tailored motivation mechanism, or bonus, may be required.

The character of this contract type means that it is not necessary to link in the cost element as part of any target mechanism, since the cost risk of the contract is naturally allocated to the contractor. However, the danger with this arrangement is that the supplier or contractor focuses on saving their own costs to the detriment of the client's schedule and operability costs.

Shared risk contracts lack any meaningful incentive, and it was for this reason that the target cost contract was developed. As explained, in Chapter 4, Contract Categories, the target cost contract is already structured with a built-in cost incentive. It was wrong to abandon it in the early days of its use. By making the incentive attractive enough to align the objectives of the supplier or contractor, with those of the client, through the sharing of savings, it can be made to work effectively. In addition, a supplier or contractor needs to be selected whose behaviour is compatible with a partnering approach, and is motivated to make it work for the mutual benefit of the contracting parties. When it is combined with other targets (schedule and operability) it can provide a powerful mechanism for achieving a successful project. Some clients link safety targets to the target cost contract incentive scheme. However, I believe that safety incentive schemes are best structured as independent stand alone schemes. This allows the contractor to focus on safety regardless of the contractual arrangements. Against this argument is the fact that the contractor could achieve cost and time targets at the expense of safety. The details of how the savings are shared between the contracting parties are described below.

The client managed risk (reimbursable) contract also lacks any natural incentive. In its purest form of a percentage fee (which should not be used!) the motivation is opposed to the client's objectives. In fact, it encourages the supplier or contractor to prolong the work for their own benefit. However, as already indicated (in Chapter 4) some slight incentive is introduced by fixing the fee. Consequently, this contract category is in most need of an effective bonus scheme. This is discussed below.

Target cost incentives

The shared risk (target cost) contract incentive mechanism is shown in Figure 12.3. It illustrates how deviations from project targets can split any gains or losses between the client and contractor. The horizontal X axis is the budget or cost axis with the target cost at the origin. The vertical Y axis is the gain or loss that is to be shared between the contracting parties. The calculation line for determining the gain or loss is set at 45 degrees, and any gains or losses from the cost target has, typically, been split between client and contractor on a 50/50 sharing basis. This is usually referred to as a Gainshare/Painshare Principle. In this illustration there is only a cap on the contractor's potential loss, but there is no reason why the client cannot impose a cap on the potential gain. If the calculation split is, say, 60 per cent to the owner and 40 per cent to the supplier or contractor, then it may be perceived as inequitable and produce some negative reactions. Conversely, the contractor may be attracted by a 60 per gain split with only 40 per cent to the client or owner. However, the loss allocation would, in all likelihood, have to be reversed.

A similar diagram is required for the schedule target. In this case the X axis is the schedule scale with the target completion date set at the origin. The Y axis is then set to indicate the gain to be achieved. If the gain paid for schedule improvement is defined on a daily basis, the 45 degree calculation line will be as illustrated for the cost target diagram. However, if the schedule gain is paid

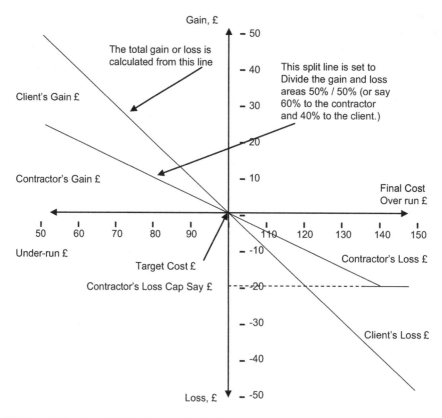

Figure 12.3   Gainshare/painshare principle

on a weekly basis then, instead of being a straight line, the diagram takes on a stepped arrangement. The total gain or loss is then the summation of both the cost and schedule incentive diagrams.

The format of this diagram is useful since the area allocated for the contractor's gain or loss can be subdivided to allow for additional contractors and suppliers in an alliance. The area would be subdivided in proportion to the risk that each alliance member was taking. Under these circumstances, it is vital that the members of the alliance have more to gain by working together, as a team, than as individuals.

The cost diagram can be simplified for a single contractor by eliminating the split line and adjusting the vertical Y axis to reflect the scale of the actual payments that the client is willing to pay – in effect a bonus scheme.

BONUS INCENTIVE SCHEMES

A well-structured bonus scheme should be self-financing and, ideally, provide additional benefit to the client. Thus, the additional revenue stream generated by early delivery or completion must be greater than any bonus earned by the supplier or contractor. Consequently, the emphasis of a bonus scheme is a target completion date. However, early completion must not be achieved by excessive expenditure or by compromising the quality of the workmanship. As a result, targets need to be defined in terms of the conventional project objectives of Cost, Time and Quality/Operability, and need to be interlinked.

If a bonus is earned by one of the targets, but to the detriment of the other two objectives, then the incentive formula should negate any bonus earned.

A bonus scheme needs to be set up, as a contract, independent of any primary contract under the sole control and adjudication of the client and it should be offered as a voluntary arrangement. If it is used on a client managed risk contract then the bonus should be structured so that the supplier or contractor cannot make a net loss from the scheme. This is what distinguishes a bonus scheme from a shared risk target cost contract.

The contract for a bonus scheme needs to be available as soon as the winning tender has been identified. It needs to be presented to the proposed supplier or contractor before the primary contract is signed so as to maintain the competitive pressure. The scheme must be perceived as equitable and the targets must be credible.

I will always remember being asked by a client if the schedule we had proposed was the best we could achieve. Naturally, recognizing an opportunity to gain a competitive advantage, our project team worked hard for a few days and we responded that we could reduce the schedule from, say, 52 weeks to 48 weeks. 'Well done,' said the client, 'we will offer you a bonus if you can beat 46 weeks.' My initial reaction was that we had been cheated – we had taken all of the fat out of the schedule and now we had to aim for an even shorter schedule! It took a moment or two for the penny to drop that that was just the point, and in any case we had nothing to lose. The incentive worked, and we did beat the client's target and earned a bonus.

Again the schedule is along the X axis with the target completion at the origin and the slope of the calculation line is set to reflect the actual bonus paid on the Y axis. It is also possible to provide a further incentive by including a step change in the schedule calculation line after, say, a 2-week improvement, with any further improvement paid at an increased daily rate, see Figure 12.4.

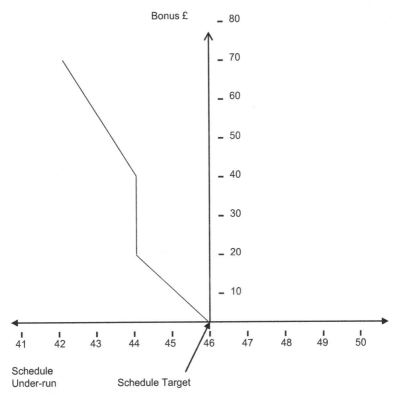

Figure 12.4   Bonus payment scheme

For example, with a target mechanical completion date of 46 weeks a bonus of, say, £2000 per day could become payable for each working day achieved ahead of the target up to a maximum of 10 working days. At that defined date an additional £20 000 could be payable and any schedule improvement earlier than the defined date would be payable at, say, £3000 per day. Any cost-saving bonus would be added to the schedule bonus within an overall bonus limit of £100 000.

Completion needs to be defined in terms of the scope in the primary contract document. For materials and equipment, the purchases will need to have been inspected and tested and to be usable or operable. However, despite delivery of the main items of an order, there still needs to be an incentive to deliver documentation and spares or other smaller items that are parts of that order. Consequently, payment of any bonus should be contingent on completing these lower-profile items satisfactorily.

Completion for a contract should relate to the Mechanical Completion Date, or beneficial use of a facility, as defined in the primary contract document. In this context, mechanical completion is defined as a fully tested facility but before live fluids or other hazardous materials have to be used in order to operate a facility. All activities up to mechanical completion are the responsibility of the contractor. Subsequent to mechanical completion the client's operations personnel are in control.

In order to discourage changes, the schedule target should be defined as inclusive of all changes, including authorised changes. This provides an added incentive. Where the client is managing the risks and reimbursing the costs, the target cost needs to be the cost forecast by the contractor at contract award. However, an overrun allowance, for the authorised changes, can also be included for in the cost target, before any negative effects take place.

The total bonus would be the sum of the schedule gain and (say) 50 per cent of the cost gain. Similarly, any schedule gain or bonus paid is reduced by the agreed share (50 per cent) of any cost overrun. As a further incentive, any cost gain or bonus should only be paid provided the target completion date is achieved.

It is vital that quality is linked into the bonus mechanism by measuring the ease of achieving a fully operating facility. Any bonus should be reduced (at a higher daily rate) for any interruption or delay to the commissioning process due to deficiencies in the contractor's work. Since setting any system or process to work is fraught with problems it is reasonable to allow 2, or maybe 3, days grace for solving problems before a reduction in the bonus is implemented. Since numerous interruptions to the start-up process would be an indication of poor workmanship by the contractor, the totting up of any delays must be cumulative. Measuring the successful achievement of quality, in terms of reduced maintenance, is more difficult or impractical due to the longer timescales involved.

The commercial arrangements need definition with an overall limit on any bonus payable. Payment should only take place after all other issues and payments have taken place. The contractor should be responsible for any taxes since a bonus is, in effect, additional profit to them.

THE NEGATIVES OF COST INCENTIVES

Having suggested that the cost, time, quality objectives should form part of an integrated incentive mechanism there are, perhaps, more powerful arguments for omitting the cost saving part of an incentive scheme.

The dilemma is that each of the cost, time, and quality elements can work against each other. Purchasing cheap materials to achieve cost targets can have a significant impact on operability. (See example of a compressor vibration trip system in Chapter 7). Similarly, the rush to achieve schedule targets can have an impact on the quality of workmanship and, hence, operability. By contrast, the attention to detail and workmanship during the execution phase is likely to delay progress, with the consequent delay in the client achieving beneficial occupation or operation. Further, the focus on quality and operability may well cost more in the short term but save on maintenance and operability problems in the longer term. Unfortunately, the client project manager or client buyer is judged on short-term targets of cost and schedule and behaves accordingly. It is the user who can be left with the longer-term quality problems.

I have previously stated that how people are rewarded determines their behaviour. As a result, money (cost saving) is a more powerful driving force than time. Consequently, a cost incentive target will take precedence over a schedule incentive target unless the schedule savings are seen to generate a larger proportion of any financial bonus. An article in *The Observer*[2] provided an example of how incentives change behaviour:

> *The surprise 40 per cent rise in GPs' salaries since 2002–03 is…the outcome of a crude performance-management system that allots points for targets (for instance, seeing patients within 48 hours) and pounds for points. Last year, the first year of the new contracts, GP practices were expected to tot up 700 points out of 1050. People with sufficient incentives almost always meet their targets, usually at the expense of unmeasured aspects (try booking an appointment with your doctor more than two weeks ahead). In fact the average score hit 950. This year it is higher.*

Further, it is easier to identify costs savings as discrete stand alone problems for individuals to solve. Whereas, schedule improvements are more complex and require a concerted and integrated team effort.

Perhaps most importantly, 80–85 per cent of the cost generating decisions have been determined in the early phases. The cost influence curve, Figure 12.5, shows that by the time the project is performing detailed design and the purchasing of materials and equipment, only 10 per cent of project costs can be reduced. This 10 per cent relates to good design practice and effective negotiation of the commercial arrangements. Since the scope of purchases is fixed the only opportunity for additional cost reduction is, therefore, in the quality element. For example; in Chapter 11, Evaluating the Tenders, I claimed that, 'The cheapest machine is almost always the machine with the lowest operating efficiency and highest maintenance cost.' Thus reducing cost is in direct opposition to the long-term operability and maintenance objectives of the owner.

In general, in can be argued that few owner clients perform sufficient contracts with incentive mechanisms for them to gain the requisite experience to determine the best ways of motivating contractors. Contractors, on the other hand, gain a much wider experience of different mechanisms with different clients and learn how to manipulate them to their advantage. Contractors who gain knowledge of the contracting philosophies of individual clients will tend to inflate estimates and schedules when targets are being established. This particularly applies to target cost contracts. I know of a leading offshore contractor who, in the late 90s, used probabilistic analysis, during the final tender negotiations, to convince the client that a more 'realistic' (later) completion date should be defined in the contract, and they succeeded, and maybe they were right.

PAYMENT INCENTIVE CASES

The following two cases[3] are examples of using payment terms to act as the incentive mechanism. They are based on real projects, but have been heavily adapted to act as teaching exercises. In

---

2   'How Labour turned the UK into a Soviet tractor.' By Simon Caulkin. A management article in the Business & Media section of *The Observer*, 23rd April, 2006.

3   These scenarios and solutions were developed with Stephen Carver who was Project Manager of Chay Blyth's British Clipper's project.

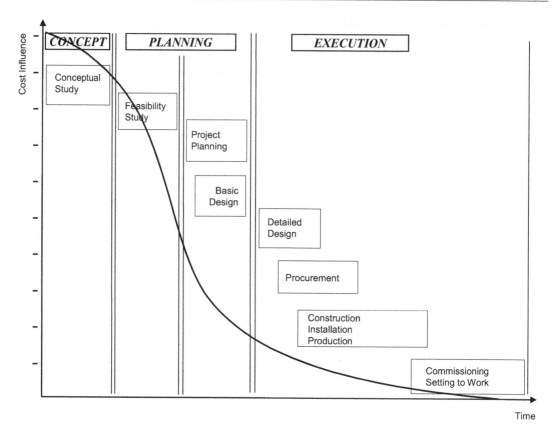

Figure 12.5   Cost influence curve

determining the payment terms, to complement an appropriate contract strategy, the emphasis should be on using time as the motivator. Further, identify the advantages and disadvantages of alternative arrangements from both buyer/client and seller/contractor perspectives. Suggested solutions are discussed afterwards.

Floating Asset Restaurant

A hotel and catering company has been awarded a license by the London Docklands Development Agency to operate a very upmarket floating restaurant alongside a recently renovated quay. An old cargo vessel has been purchased that the client intends to convert into the proposed restaurant. A design company has been appointed and they have prepared preliminary layouts. The design company has pointed out that until strip-out work on the vessel is started the full scope of work cannot be assessed. The strip-out has to be undertaken in the dry dock of a shipyard but, once started, it will be impossible to move the vessel again until the conversion is complete. A suitable contract must be developed for the conversion work.

(This scenario could, in a similar way, be applied to property renovation.)

Slick Oil Inc.

A major oil company has had a massive blowout in one of its oil wells. Oil is gushing out and polluting a major river in an environmentally sensitive area. Only one or two companies in the area have equipment suitable to choke the well. The oil company's operational director has already had brief discussions with one of the companies and is now on the telephone to the second contractor. What contract and payment terms should be negotiated?

(This scenario could be applied generically to any spillage of hazardous materials and equally to any disaster situation. For example, the crippling of BP's Thunder Horse oil platform in the Gulf of Mexico by hurricane Dennis in 2005. Part of the leg system, of this $2 billion project – the largest platform in the Gulf of Mexico – was punctured by a stray two tonne anchor that should have been lashed down in the hurricane evacuation, but was not secured).

Discussion of cases

Floating Asset Restaurant

The key risk to this project is that once the vessel is in the dry dock the shipyard will immediately open up the vessel for inspection. Once hull plates have been removed, and the vessel can no longer be moved, work will slow down. Since the project duration cannot be defined, the owner is stuck with a vessel in dry dock on a daily rental rate. Consequently, a separate agreement must be made for a fixed price for welding on hull plates to make the vessel watertight. It is important that the shipyard understands that the client has the option to have the vessel towed away at a moments notice. Knowledge of the marketplace is essential. This is a buyer's market scenario – shipyards need work, so alternatives are available.

The next problem is the scope, which is clearly a client risk. The definition of the new scope of work cannot be finalized until strip-out work has been completed. However, the strip-out scope cannot be finalized until all defective materials have been removed and hidden work exposed. An independent survey is, therefore, essential. It is then not unreasonable to ask the shipyard to take the risk for the visible and defined work to be torn out. This is, after all, their expertise and a fixed price can be agreed. However, the subsequent strip-out work that is exposed is a client risk. Consequently, rates for this type of work need to be defined and scheduled so that any strip-out work that will expose additional work is performed first. Further, the development of the final design details needs to be carried out in parallel with the additional strip-out. This should avoid some of the possible schedule delays.

With a clear vessel (site) and a defined scope the shipyard can be asked to take the risk for the new build. Again this is their expertise and they are in the best position to control the risks involved. Whilst the client would like to have payment terms of 100 per cent when the vessel is seaworthy, it is unlikely that the shipyard would be in a position to carry this risk. The cash flow required for the wages bill is likely to be too onerous. Consequently, regular payments will have to be agreed, probably monthly, and provided the progress of the work is satisfactory. However, the client must avoid overpaying as already discussed. The client is in a slightly better position on this project since the support facilities required for the project exist as part of the dockyard infrastructure. Therefore, no down payment will be necessary. Obviously, the contract terms need to be negotiated during the strip-out work. The client must a) be in a position to remove the vessel to another location and, b) maintain a competitive climate to the negotiations.

Additional security against overpayment needs to be devised and an incentive structured into the payment terms. This can be achieved by extracting the overhead and profit elements from the payments due. The overhead and profit would then only be paid when key schedule (and easily definable, visible or tangible) milestones had been achieved – for example, erection of mast(s), functioning engine and rigging completed. The money motivator has thus been used to become a schedule driver.

Slick Oil Inc.

The major risks are environmental. The issues are urgency, duration of the work and availability of suitable contractors; all of which will affect the public's perception of the company's reputation. However, the speed with which action is taken will be the key to this PR perception.

Urgency of mobilization means that there may be little choice but to use the local contractors. The two contractors most likely see the situation as an opportunity to make money. It is an opportunity that cannot be missed. Perhaps the oil company's operational director has already detected this in their telephone conversation with the first company. Maybe the first contractor decided to take advantage of the situation and quoted an exorbitant figure, and this is why the oil company is now talking to the second contractor. Consequently, the oil company will be seeking a solution that satisfies the contractor's desire to make money, and their desire to get the work done quickly.

The proposal that could satisfy this requirement is an offer to pay the contractor their standard rates for 'normal' work, with a fixed sum for overheads, together with an incentive scheme. The scheme will be based on a direct application of the one described earlier in the chapter. However, the two tasks, capping the well and cleaning up, should be treated independently. If the oil flow is stopped within an hour a very high bonus is paid, and the longer it takes the bonus is reduced on an hourly basis. If it takes longer than a day, a step change reduction is applied and the bonus is paid at a lower level. Ultimately, when the time taken is unacceptable, the bonus is reduced to zero and only the standard rate is applied. The bonus offered must be sufficiently high to make aiming for a short duration worthwhile. Otherwise the contractor may decide that it is more profitable to spin out the work. However, the fixed overhead should make this option less attractive.

It is worth the oil company making the bonus for stopping the oil flow really attractive. Obviously, the quicker this work is done the less work and cost is involved in the clean-up.

The clean-up work would be based on a similar formula but with a daily, rather than an hourly, time schedule. The contract must allow for the oil company to separate out some of the work and award it to a second contractor if necessary. In addition, it must contain a break clause allowing the oil company to terminate the contract at any time.

The intelligent contractor should realize that there is an opportunity to do the oil company a favour – build relationships – and, as a result, be given the opportunity for additional work in the future. Consequently, there is a good chance that they would accept this proposal.

Alternatively, the client could suggest that the two contractors work together and negotiate a joint approach. This might be attractive to the contractors on the basis that half a cake is better than none.

13 *Other Influences*

The political system, the cultural norms and the state of development of a country all influence the manner in which the purchasing of goods and services deviates from the theoretical ethical process. Nevertheless, it is up to the project manager to set and maintain the ethical standards on a project.

Of course, corruption is most likely when buying a standard product, where price and delivery are the only factors to be considered in making the buying decision. There are two reasons for this. Firstly, if the technical merits of the tenderer's proposals differ, either because of the solutions offered or the standard of design, then to prefer the wrong tender could result in the purchase of inferior materials or equipment. A more serious matter and one that is less likely to occur than if all the offerings were technically equal (for example with standard materials). Secondly, if there are both technical and commercial aspects to the purchase, then the decision is a joint one between the technical and commercial people. Consequently, any funny business must involve collusion.

POLITICS

Naturally government policy, the United Nations, the World Bank and other lending banks, the European Union Directives and other official organizations' rules will be primary determinants in awarding contracts, whether for goods or services.

Dealing with contracts at the highest levels of government can be a hazardous business:

[1]*ONE of Britain's richest men has launched a record £200m legal claim against the French oil company Elf in the high court.*

Nadhmi Auchi, an Iraqi-born billionaire, started the action after being convicted of corruption in France, an outcome he insists was caused by the actions of Elf.

A French court will this week rule on an appeal launched by Auchi against his conviction but the billionaire claims that his relationship with the multinational oil company has caused him huge financial damage.

The case centres on a deal involving Elf, then state owned, to buy a Kuwaiti oil refinery from Auchi during the first Gulf war.

---

1   'Tycoon launches £200m claim in Elf bribery case.' by Robert Winnett. *The Sunday Times*, January 28, 2007.

Elf asked Auchi to buy the refinery – which the Kuwaiti government had to sell within days after being invaded by Iraq – and then sell it on to the French oil company after getting approval for the deal from the EU.

French investigators claim that Auchi paid bribes to Elf. The billionaire says any payments were legal and proper and were made at the behest of a French Government entity. He was personally thanked by President Mitterrand for his role in the deal.

Organizations representing government tend to change their minds or change the rules. On a project in Algeria we discussed (with the owner client) all the taxes and duties that would have to be paid. They suggested that we talk to the tax authorities, which we did. We then followed the rules and regulations that they said applied, only to be presented with a substantial fine towards the end of the project. Naturally we protested to the tax authority, pointing out that we had followed everything that they had advised us to do. They agreed with us but said, 'We were wrong,' and further, 'because you have not paid the right duty on time, not only do you have to pay the duty, but we also have to fine you an additional 100 per cent!'

In the very early days of the development of the North Sea oil industry the company I was working for was the leading contender for a major design contract. However, we were told by the client, at the very last moment, that we could not be awarded the contract because we were owned by an American company. That Christmas we gave a 51 per cent present to a British company, and the new joint venture signed the contract.

In South Africa a Black Economic Empowerment (also known as BEE) policy of preferential procurement takes place whereby all contracts are awarded to 'black' businesses. The average size of contracts handled by black economic empowerment companies is £5m and some companies can handle projects up to £55m. However, it is acceptable to facilitate joint ventures and partnerships with established non-compliant companies. So that 'white' companies are only likely to be awarded contracts when the work involved is too big for the local black companies. In one advertised request[2] for Expressions of Interest, 10 points were 'allocated to the black economic empowerment credentials of the bidder measured in terms of the Codes of Good Practice issued by the National Department of Trade and Industry of the Government... .' The purpose of black economic empowerment procurement is to create an economy where all businesses compete on an equal footing regardless of the status of the owners.

[3]Over the next decade or so, the playing fields are going to be levelled and companies won't be able to rely on their BEE status to secure business. Companies will be looking for service providers that can deliver, full stop. Instead of waiting for this to happen, we are focusing on the future, now. Delivery is the number one attribute we strive for, along with honesty, commitment and passion. Our BEE status may open doors for us, but it is not what will keep the doors open. Once you have been given an opportunity you better be able to prove that you can do the job.

FACILITATION PAYMENTS

On the whole it is the relationship that matters most. People like to do business with people who they know and like and who understand their way of doing business. All things being equal it is the quality of the relationship that will influence the decision-making process. To quote from the Holyrood

2   *Financial Times*, May, 2006.
3   Karel Bezuidenhout, CEO Loropo Industries in an article about project management – 'always a step ahead.' South African magazine *CEO – Celebrating Excellence in Organisations*, Vol 5 No 9, 2006

Inquiry, 'Although their tender bid was not the lowest one it was crucial that a company that the client and Design team felt comfortable with was selected for this most crucial of roles in the Project.'

Corporate entertainment is not offered or accepted as a bribe, but as a means of promoting good relations and securing access. It is difficult to refuse to see a salesperson if you were at a rugby match with them last month. All the same, there are limits which are difficult to define and, perhaps, easy to cross. Consequently, it is when people prey on our more basic instincts that one should be wary.

The predicament starts with gratuities paid as appreciation for services already rendered. They are paid after the performance of the service, but, in a continuing relationship, it is also in expectation of something better in the future. This escalates into a generous gratuity in order to influence the relationship and is on the cusp of changing from appreciation of a service to a bribe to obtain some additional service other than the norm. In the days of traditional restaurant cars on trains my mother always sought out the attendant and 'tipped' him to let her know when the meal was ready to be served. At the appropriate time the attendant would come and tell her, and we would be shown to the seats especially reserved for us. Thus, is it still a gratuity if a crisp note is slipped into the hand of the Maitre d'Hôtel, before the event, in order to reserve that prime table in the window where you want to celebrate your wedding anniversary? Surely it has become a bribe once a payment of this nature is made in advance. In the 1980s in Indonesia it was often necessary to pay a bribe, euphemistically called a Facilitating Payment in order to pay a bill.

The Italians could be quite subtle in the way that they crossed the line and bought clients' goodwill with facilitation payments, or should I say bribes. Overseas missions are regarded as perks to personnel in less developed countries, and are often apportioned out as rewards or bonuses for good work. In order to reduce foreign exchange costs a per diem is paid rather than expenses. Knowing this, the Italian vendor pleads urgent need for a client specialist to visit the vendor for every little problem. The client visitors are met, transported, dined and accommodated, and so on. The vendor then quietly picks up the bills. UK vendors, on the other hand, always preferred to visit the client and see the sights. On the rare occasions when the client is invited to the UK they are booked into expensive hotels and left to pay their own bills.

BRIBERY AND CORRUPTION

In business transactions the equivalent of the gratuity is the Christmas gift made as a goodwill gesture for the orders or contracts received during the year. But it is also in expectation of enhancing relationships, and influencing the recipient to look more favourably on the giver. For project, and particularly purchasing, personnel this can easily get out of control and, by the 1970s in the UK, this culture was out of control.

I remember one contractor presenting me, as project manager, with a whole series of extras that I could not understand. The contractor explained to me in a hushed voice that these were for resurfacing the private access road to Mr G's (my boss's) mother's house. I told the contractor, firmly, that I would not authorize them and that they should present the bill to the person who had asked for the work. Other experiences demonstrated that he was, and had been, benefiting from 'favours' from contractors over a number of years – I do not believe that he was ever brought to book.

The Aberdeen based company, Aggreko, that rents out power generators has become one of the major suppliers of energy to Kenya, Uganda and Rwanda[4]. Their chief executive said '...that the African governments, contrary to popular perception, were excellent to do business with. "We have more bad debts and more stolen equipment from within the M25 than we do from Africa."'

---

4   'Aggreko powers into bid for Beijing Olympics' by Harry Wallop. *The Daily Telegraph*, March 9,2007

Bribery seems to be endemic in most countries. There are certainly cultures where people, who have achieved positions of power, are expected to use their positions and influence for their own benefit.

As a contractor, buying goods and services and letting contracts in an Afro-Caribbean environment, you will need a local partner. You will also need the services of someone of influence – a Sponsor, to help you win the main contract work. The sponsor will find a local partner for you who will have a 25 per cent stake in the project. For, say, a US$250m project their loss will be capped at US$7m and their profit capped at US$5m. The local partner's involvement will be restricted to providing one site representative (who, incidentally, will do absolutely nothing for the duration of the project). The sponsor will not require any commission or fee for their advice and influence. However, the sponsor's wife will own a services company and her company will want the opportunity to match the lowest price of any tenders submitted. This process will operate for up to an overall limit of US$5m worth of orders. Accordingly, the wife's services company is presented with the lowest tender in order to match the price. The services company then says to the local company, 'You will not win this contract unless you can reduce your price by 5 per cent.' The local company agrees. The wife's services company matches the tender price and passes on the work to the local company at the 5 per cent lower price. Whilst one might be critical of this process it has to be acknowledged that when operated 'properly' projects get completed on time. The alternative is for governments to spend hundreds of thousands and millions of dollars for no results.

Using ones position for ones own benefit might be seen by the, possibly, more arrogant business person in the West as only applying to less developed societies. I believe it is alive and well in, for example, Italy and other parts of Europe.

> [5]*Corporate Germany was in the dock last week when a former Volkswagen boss was convicted of bribery. [The] former personnel director...shuffled his way across a cobbled yard to a courthouse in Brunswick to confess to bribery and corruption on a grand scale.*
>
> *At the precise moment Hartz stood before the court, in Munich the supervisory board chairman of the engineering giant Siemens was...apologising for another corruption scandal – the worst in German corporate history. ...*
>
> *It was the moment Germans began to ask themselves who else is guilty. How deeply ingrained is institutionalised corruption? Who let them get away with it for so long?...*
>
> *'In the past, cases of corruption in a company were more frequently covered up or played down. Nowadays there is more public awareness and people are more sensitive to the issue.' said Peter von Blomberg of Transparancy International, a group that fights corruption around the world.*

As indicated above, what might be argued as simple facilitation payments to low paid officials can easily escalate to bribery and corruption on a grand scale. Consequently, it was a breath of fresh air when Exxon wrote to their suppliers and contractors in the early 1980s saying in effect, 'Please do not send us gifts at Xmas. However we are happy to exchange diaries and calendars.'

This fresh approach, together with awareness of the problem and better procedures has, hopefully, eliminated the worst excesses at the lower ranks of an organization; however, it seems to be worse than ever at senior levels.

---

5   'Dirty Rotten Business' by Michael Eoodhead. *The Sunday Times*, January 28, 2007.

The Dutch anti-cartel watchdog the Nederlandse Mededingingsautoriteit (Nma) fined 344 civil engineering and infrastructure contractors £162 million for price fixing and market-sharing behaviour.[6]

The initial Nma investigation which resulted in fines totalling € 100 million, revealed that contractors, by coordinating their bids, had increased the cost of building a major tunnel near Schipol airport by 8 per cent. ...

Britain's OFT [Office of Fair Trading] is watching the Nma investigation, in particular its success in encouraging 140 companies to act as whistleblowers. Companies that voluntarily provided information will have their fines reduced by up to 50 per cent.

The Office of Fair Trading said that there was rampant bid rigging among construction companies tendering for public sector contracts.[7]

A large amount of evidence of collusion...was provided by whistleblowers hoping to escape or minimise punishment by cooperating with the OFT.

...some building firms had paid money to rival companies to persuade them to stay out of bidding for contracts. In other cases, construction companies were operating a so-called 'slate' system, in effect secretly dividing up contracts among themselves.

'We consider such arrangements are likely to meet the dishonesty test as set in the criminal jurisprudence,' [said the director of cartel investigations at the OFT].

He said that even cases of 'cover bidding', although not regarded as particularly sinister by many in the industry, still breached the Companies Act. Cover bidding is the common practice of firms putting in an uncompetitive bid to ensure that they stay on the tender list for future contracts rather than with any intention of winning business.

...The Competition Act 1998 prohibits agreements, practices and conduct that have a damaging effect on competition in Britain.

The Taiwanese Chinese were very easy to deal with in the 1980s, easier than most UK contractors. Where a continuing relationship existed they expected to recoup a loss on one contract by charging a bit more on the next. However:

[8]The sad fact is that the [mainland] Chinese system today is almost incompatible with honesty – almost everybody is at least a little bit dirty.'... here is some of his advice 'Once you get below the level of the big multinationals doing large deals, China becomes a swamp.'...To be fair to McGregor, he says there are honest Chinese firms and honest officials, high and low, but his 20 pages on the details of official and business corruption are followed by barely one on how to function honestly ... Although McGregor says occasionally that not all Chinese are corrupt and many companies 'insist on high ethical standards', overall he warns businessmen to keep a tight hand on their wallets.

In Korea, copies of competing contractors' tenders were easily available to the competitors. In Thailand, matters were not as bad, but one could not rely on anything being as it seemed. The

---

6   'Dutch to name 344 price-fixing firms.' *The Times*, February 21, 2005.
7   'Rigged building bids rob public sector.', by Patrick Hosking and Rajeev Syal. *The Times*, December 2, 2005.
8   Extract from 'How not to lose your shirt in China' by Jonathan Mirsky. Book review of 'One Billion Customers: Lessons From the Front Lines of Doing Business in China' by James McGregor. ISBN 1857883586. *The Spectator*, 10 December, 2005.

following project example in Thailand[9] describes more convoluted relationships than the more straightforward Afro-Caribbean example above.

Again the client, (in this case an owner client) had a local (40 per cent participation) partner. However, the local partner formed a joint venture with a Korean company to tender for the construction work that was part of a main contractor's (Davy McKee) tender for the whole project. The project manager soon realized that anything that was said to the local Thai partner was passed on to the company (Davy McKee) tendering for the main contract. Consequently, the project manager made sure that nothing about any of the tenders was ever discussed with the local partner until his final report and recommendation.

The project manager received a few messages from companies that were tendering for equipment saying that Davy McKee believed that they had more influence on the awarding of contracts than the project manager. As a result, when he recommended that a contractor other than Davy McKee be awarded the main contract, all hell let loose.

By containing information the project manager was able to obtain the agreement of the project's joint venture board (ICI and local partner), because the local partner's representatives were caught by surprise, assuming that the construction contract was in the bag. The local company then pressurized the ICI business manager to instruct the project manager to re-tender the contract on the spurious grounds that Davy McKee had not been treated fairly. They did not believe that an employee could refuse an order (in Thailand this would be unthinkable). The project manager refused and resigned from the project. The ICI Engineering Director said if Roy Whittaker was not doing the project, ICI Engineering was not doing the project. Then the ICI Deputy Chairman said that if Roy Whittaker was not doing the project, then ICI would not do the project, even if it had already been sanctioned by ICI. Now that is management support! So the project was cancelled.

In 1985 the Al Yamamah (meaning the dove) arms deal, worth £43bn, was signed between the British and Saudi Arabian Governments. The contract included a £60m fund to pay various expenses, such as: buying property and limousines, providing security, arranging visits to London/Cannes/Milan, paying TV bills and so on. This fund was run by a travel agent acting for BAE Systems. BAE consistently denied any wrongdoing and was quoted as saying on BBC Newsnight:

> On December 14th 2006 the Attorney General announced that the Director of the Serious Fraud Office had decided to discontinue that part of the investigation which related to the Kingdom of Saudi Arabia.

This was a contract with a culture where 'gift giving' is an accepted part of doing business. However, there was a culture clash with the British where 'gifts and personal expenses' are seen as bribes, and as such, against the law.

The fundamental issue is that if the bribe is big enough, 'We will cancel our latest £10 billion defence contract for Eurofighters unless you drop your investigation into alleged bribery,' even governments will succumb. The Attorney General's[10] explanation was shocking, 'It has been necessary to balance the rule of law against the wider public interest.' What kind of message (from the top) does it give the business community? It says to me that there are issues that are regarded as more important, such as winning the next contract or that the client can be persuaded with the right influence. The issue continues to rumble on:

> [11]The UK is covertly trying to oust the head of the world's main anti-bribery watchdog to prevent criticism of ministers and Britain's biggest arms company, BAE,...

---

9   Provided by Roy Whittaker, ICI Project Manager of the project.

10   System failure', by James Harding, Business Editor, *The Times*, December 15, 2006.

11   Headline article 'UK tries to sabotage BAE bribes inquiry', by David Leigh and Rob Evans, *The Guardian*, April 24, 2007.

The effort to remove [him] comes as his organization has stepped up its investigation into the British government's decision to kill off a major inquiry into allegations that BAE paid massive bribes to land Saudi arms deals.

British diplomats are seeking to remove…a Swiss legal expert who chairs the anti-corruption watchdog of the Organisation for Economic Cooperation and Development (OECD) claiming he is too outspoken.

…A source at the OECD added, 'The UK's representatives were sent to Paris to emasculate the [watchdog] and ensure they did not say anything publicly. They failed and were not pleased. They behaved in a manner that would not have been out of place in a boxing ring.'

…It also rebuked the UK for failing to keep its promise to modernise its inadequate corruption laws, under which no one has yet been prosecuted.

The UK also faces a legal challenge in London. Two campaign groups, the anti-corruption group The Corner House and the Campaign Against the Arms Trade, filed detailed pleadings last week alleging that Britain had broken the treaty banning corrupt payments by companies to foreign politicians and officials.

CONTROLLING THE MONEY FLOW

A project manager working in one of the states of the former Soviet Union summed matters up correctly, 'There is no doubt that there are countries where the "brown bag culture" exists. Consequently, the project manager needs to be involved in the purchasing process if they are to stop the money running out of the door.'

Arabs in general, and Egyptians in particular, are naturally of a suspicious nature[12]. The chief executive of a company providing engineering and procurement services in Egypt believed that sellers of goods and services employed the most capable and expensive experts to deceive buyers. Sellers hire, reward and promote their personnel on the basis of the relationship they have with someone in the potential buyer's employment. Consequently, this CEO operated on the basis that persons involved in the buying and selling of goods and services may be dishonest or, if not, they were potentially corruptible.

The Ford Motor Company, for example, would rotate buyers from one specialist component buying role to a different specialist component buying role in order to prevent too cosy a relationship from developing.

As a result, this Egyptian CEO believed it was essential to have a well defined and detailed process for controlling the procurement of goods and services. Key components of this process being;

1.   The management of a Technical Tender Opening Committee.

    and

2.   A Commercial Tender Opening Committee.

In his book 'Project Management in the Process Industries'[13] Roy Whittaker states that the project manager must resist, to the point of resignation, pressure to behave unfairly. The price of freedom from corruption is eternal vigilance and Roy Whittaker's rules for dealing with it are:

Rule 1:   Assure staff that you are not going to preside over a corrupt organization, and that you take it personally.

---

12   Experience from Vernon T Evenson; Project Manager.
13   *Project management in the process industries*, Whittaker, R., page 164, John Wiley & Sons, 1995. ISBN 0-471-96040-3.

Rule 2:   Avoid temptation. Ensure that every order needs two signatures so that corrupt behaviour also requires collusion.

Rule 3:   Investigate every rumour diligently and make it known that you are doing so.

Rule 4:   Make it very clear to staff that the taking of bribes means instant dismissal.

He states that he never got to the bottom of any rumour but believes that he frightened a few people on the way! Where he was sure that there was something not quite right, but could not prove it, he had no compunction in calling in the offending supplier and telling them that they would get no further orders, and why.

14 *Finalizing the Deal and Delivering*

The final choice of supplier or contractor might be completely clear – the offer conforms to the required scope and specification, the delivery is acceptable and it is within budget. In these circumstances the buyer can be left to negotiate, submit the selected recommendation for approval and then finalize a contract.

NEGOTIATIONS

In discussing the evaluation and selection of the successful tender some clarifications were obtained. However, no negotiations were involved before the final selection was made and endorsed by the project manager.

It is much more likely that a series of negotiations, both technical and commercial, will be needed. The technical negotiations will take place first to ensure that the scope and specification requirements are as needed before discussing the terms of the contract. It is not uncommon for the technical people to be rigorously controlled and constrained by the commercial people during the negotiations. This is because technologists tend to give signals that are invaluable to a seller, 'Your equipment is the only item that meets the required performance criteria.' It is obvious that this places the buyer at a disadvantage in any subsequent commercial negotiations. The reason for this faux pas is not the technologists' fault. In organizations where this could occur there is a strong segregation of fiefdoms. The procurement discipline becomes protective of its expertise and tends to exclude the technical functions from getting too involved. Further, the organization as a whole does not see the necessity to train technologists in commercial skills – after all they are technologists. A project is constantly negotiating different aspects (scope and specification changes, resources, schedules and so on) and an enlightened organization will train all project personnel in negotiating skills. The more far-sighted organization will not only train the technologists in negotiating, but also give them the total responsibility of buying from specification, to delivery and installation.

It is not intended to provide a dissertation on negotiating skills. Nevertheless, knowledge of the basics and the process involved is essential in order to buy goods and services. After that, it is a matter of practice, practice and practice.

Early books on negotiation are based mainly on personal experiences. They tend to deal with various issues in isolation without linking the issues to an overall negotiation process as a whole. It was not until the publication of Gavin Kennedy's first book 'Managing Negotiations'[1] that all the

---

1   The two other authors are: John Benson and John McMillan and the book is subtitled 'How to Get a Better Deal'. Random House, 1980. ISBN 0091415802. This book was out of print in 2007.

experiential issues were brought together in a formalized process. The following is a brief summary and explanation of the phases common to all negotiations.

- Prepare: identify the 'must haves'. Gather the facts.
- Discuss: listen and exchange information. Explore interests.
- Signal: recognize and confirm messages.
- Propose: 'if...maybe...' Try to increase the issues to be considered.
- Package: 'If you do...then I will...'
- Bargain: never concede something without getting something in return.
- Close: 'If you do that then we have a deal.'
- Agree: agree what has been agreed. Summarize and record the agreement.

In discussions, the buyer will be asking questions to understand why one tender is different from other tenders. The buyer should endeavour to get the tenderer to provide a price make-up. A price can be broken down in any way that is requested, whereas a price make-up helps to determine the actual costs. It helps identify where a particular element, say packaging, is significantly different from the expected norms. Perhaps the tenderer has interpreted the specification or the contract terms differently from that intended. Consequently, seeing a different or more severe risk the tenderer may have included additional monies to cover for the risk. By explaining what was intended the buyer can legitimately get the price reduced.

Thus, it would also be a mistake to conclude the negotiations with, 'If you reduce your price by 10 per cent then we will give you the order.' In these circumstances the parties' objectives will only be the same up to the signing of the contract. There is a right price for a job or contract. As it is, competitive bidding forces suppliers and contractors to cut margins. So that by forcing a price reduction without a justifiable reason (a reduction in scope or risk) the supplier or contractor will seek to recover the monies, by whatever means, during the contract execution. The type of behaviour generated by unreasonably squeezing a price is usually to the detriment of the project.

Having said this, there are cultures where, having satisfied all the requirements of the tendering process, the last two or three suppliers in the race are asked to submit their Best and Final Offer. This approach is generated by a fear that, maybe, one is overpaying. Whilst this can produce better short-term deals for the client, in the longer term it has put many suppliers out of business, to the detriment of the industry as a whole.

A negotiated solution should be developed that meets the needs of both client and supplier or contractor, independent of trust. A negotiated deal should not have to rely on the two parties trusting each other. The individuals concerned may move departments, get transferred or even leave their respective companies. A new person taking over may view matters quite differently. It would be no good saying, 'But that's how we understood the deal – and we trusted each other to fulfil it.'

FINALIZING THE DEAL

Before the order or contract can be issued the appropriate approvals must be obtained. The project manager's approval is, in reality, an auditing and rubber-stamping exercise. Control should have been exercised when the enquiry was prepared and the tender list agreed and at this stage everyone will be anxious to 'get on with it'. Never the less, the project manager will want to make sure that all the work has been performed correctly. This will involve reviewing the documentation for any blank boxes in the standard forms that might indicate that work has been skimped.

It should be clear, that in order to prevent malpractice, the same person should not be allowed to make the final vendor selection and place the order without another person providing independent approval. The technologist recommends the required goods or services, purchasing recommend the commercial aspects, and the project controls people endorse the tender for meeting budget and schedule requirements. The project manager then approves the final selection. This final approval will, however, depend on the authority delegated to the project manager, and the delegated authority should depend on the financial size of the project. Even the most senior and experienced project manager may only be allowed to approve the placing of contracts up to £100 000 00 if the project has, say, a total value of £1m. Whereas, a less experienced project manager may be delegated authority to approve recommendations for, say £5m for a billion pound project. Above the delegated authority level the project manager will need to obtain more senior management approval, and ultimately the chief executive's approval for very big order values.

In these circumstances the project manager had better be familiar with all aspects of the recommendations in order to answer the key question, 'Why have you chosen this particular vendor?' There should be more than just the value of the order being reviewed. Since the recommendation is for a large value there will be proportionally similar risks involved. Thus, senior management will want to be reassured that the proposed vendor or contractor is responsible for their share of the appropriate risks.

Occasionally, an independent audit should be initiated by the corporate manager or director of procurement to ensure that orders are not being split into smaller packages, with separate orders for each package, in order to get around the required authority approval levels.

Placing the contract

Once all the approvals have been obtained the buyer will be under pressure to get on and place the order or contract.

Companies have their own set format for how orders and contracts are laid out. Purchasing is responsible for compiling the purchase order, which will comprise:

- the first page; the purchase order itself;
- an acknowledgement form;
- the material requisition and its attachments;
- the terms of the order, these may comprise general and special conditions;
- any additional instructions.

And:

- name and address of the client company;
- name and address of the supplier;
- the purchase order number – formed by an extension to the requisition number;
- the date and number of pages;
- the supplier's promised delivery;
- a tabulation of the item number, quantity, description of the goods, the client's cost code, a unit price of the item and a total price;
- there should also be an indication of the shipping and transport (Inco) terms and the terms of payment.

Figure 14.1 is a modified and edited M W Kellogg Purchase Order Index page. In this instance the contractor is purchasing 'for and on behalf of' an owner, since the document is headed 'client [that is, owner's] name and logo' instead of Kellogg's name and logo. The index illustrates the extent of the documentation required to detail a full description of what is to be purchased. The abbreviations SQR and SDR stand for Supplier Quality and Supplier Data Requirements respectively. Typical documents listed under Project Documents are:

- project design data;
- supplier document code listing;
- engineering handover specification;
- fabrication, erection and testing of piping;
- units of measurement;
- painting and protective coatings;
- inspection and test procedure;
- specification for general welding of pipework etc.

Other documents cover noise, instruments, electrics, emergency shut down and so on. Similarly, typical documents listed under Project Drawings are:

- general notes and abbreviations;
- symbols;
- process and instrument diagram;
- five structural steel drawings.

The requisition should be updated to reflect everything that has been discussed and agreed during the enquiry process and tender clarification meeting. However, it is possible that some minor corrections or discrepancies have occurred. Bearing in mind the requirement for Offer and Acceptance to form a contract, it is essential that an acknowledgement form is included in the purchase order package. Purchasing is responsible for ensuring that the acknowledgement is returned without any qualifications or annotations of any type.

The order package should not include attachments such as minutes of meetings or exchanges of letters. It can be particularly tempting, after discussions clarifying contracts for services, due to the desire to place a contract with the minimum of delay. By doing this, any irrelevant references to other matters or documents are incorporated into the contract and, hence, confusion occurs. Time should be taken to update scopes of work, revise contract terms or redraft other aspects.

In these days of e-commerce the order may well be transmitted electronically. However, it is important to have a mechanism that acknowledges receipt of the documents concerned. If it is necessary to fax a copy of the order, then the supplier should be requested to sign and date receipt of the fax and return it. There is, nevertheless, nothing wrong with sending the order by mail.

Purchasing is responsible for the internal distribution of priced and unpriced copies of the order or contract, in accordance with the project procedures. Typically, copies would be transmitted to the owner, the cost and scheduling or project controls group, finance and accounts, the design group and so on. Finally, the order will need to be entered on a purchase order register tabulated with the following information:

- order number and date;
- requisition number and date received;
- supplier;
- material description;
- buyer's name;
- order value.

CLIENT NAME AND LOGO Page: 1 of 2

PROJECT NAME

ELEVATED FLARE: PURCHASE ORDER No.

Purchase Order Index Page

Please use the hyperlinks (underlined) below to access the desired document. If you do not have any applications capable of handling pdf documents, please download Adobe Reader.

DOCUMENT TITLE	**DOCUMENT NUMBER**
COMMERCIAL DOCUMENTS:	
Purchase Order	
Conditions of Purchase	
Marking, Packing & Transportation Guidelines	
Bank Guarantee	

REQUISITION SPECIFICATION:	
Requisition Specification	DOCUMENT NUMBER

PURCHASE ORDER DOCUMENT TEMPLATES:	
P.O. Document Templates	

REQUISITION ATTACHMENTS:	
Manufacturing Data Report Requirements	SDR-4
Supplier Quality Surveillance Requirements	SQR-2
Supplier Data Requirements	SDR-2
List of Applicable Documentation	SDR-7
Critical Supplier Data	SDR-8
Deviation/Exception/Clarification List	SDR-10

PROJECT DOCUMENTS:	
Suppliers QA/QC Requirements (SQR-1)	DOCUMENT NUMBER
Project Design Data	DOCUMENT NUMBER
Supplier Drawing and Data Requirements (SDR-1)	DOCUMENT NUMBER

Figure 14.1 Purchase order index

Spare Parts Co-ordination Specification (SDR-14)	DOCUMENT NUMBER
Plus 19 other documents listed (see text)	

REQUISITION SPECIFIC SPECIFICATIONS:	1 – DOCUMENT NUMBER

REQUISITION SPECIFIC DATA SHEETS:	1 – DOCUMENT NUMBER

FORMS (for completion by supplier):	Supplier Data List: <u>SDR-3</u>

PROJECT DRAWINGS:	8 – DOCUMENT NUMBERS LISTED (see text)

Figure 14.1 *Continued*

Contracts will be entered on a register with the following information:

- contract number and date;
- description;
- contractor's name;
- award date;
- commitment value.

Contracts will also need a change order register to log claims for charges for extra work. For a contractor acting as a client contracts become subcontracts.

Letters of intent

Letters of intent should be avoided. A Letter of Intent worded as follows, 'It is our intention to place an order for…' places the risk with the supplier or contractor should they start work. Nevertheless, it is unprofessional to rely on the supplier or contractor to start work at their own risk. If it is necessary for work to start, then an Instruction to Proceed should be issued. However, it may be that the client intends to place a contract but, for schedule reasons, requires manufacturing capacity to be reserved. In this case, the letter should be worded accordingly and the supplier would be entitled to the costs they incur plus a reasonable profit, should the order not materialize.

EXPEDITING AND INSPECTION

The focus of this book is on the purchasing of goods and services and it is not the intention to go into a detailed explanation of other procurement functions. However, the goods or services have not been purchased successfully until they are delivered. Consequently, expediting and inspection are important parts of the process and the project manager should take an active role in assisting these functions, particularly for critical items.

On too many occasions I have heard people, from non-project focused companies, explaining that their project ran late because a supplier let them down. My usual response is that it was their fault as project manager, since one of the key roles of the project manager is to make sure that the manufacturing, installation or construction personnel have all the information, drawings and materials necessary for them to perform their activities.

Since the installation or construction personnel are the users of the materials and equipment, they are the logical people to perform an inspection before it is too late. Is what is to be delivered what they expected? They need to check that multiple shipments are scheduled in the correct order so that the items can be used as and when received. Are the overall dimensions as expected? What are the tolerances on the locating fixings? Do the connection points have the correct orientation? If matters are left to conventional processes and departmental responsibilities this involvement of the users may not take place. The project manager may, therefore, need to issue an instruction that this is what they want to happen.

The project manager should not interfere with the function of expeditors but, on the other hand, they are responsible for ensuring that the project is completed on time. Consequently, the project manager needs to be kept informed of the delivery status of materials and equipment. In fact, at some stage, the project manager will spend an inordinate amount of time reviewing material status reports with the project procurement manager. However, I found that one got a better feel for what was developing by reviewing copies of expediting reports. I would read all reports for critical materials and equipment, looking for potential problems, and only reports of matters that were not going according to plan – exception reports – for non-critical materials and equipment. The danger is spending too much time in one area and not seeing the wood for the trees. Consequently, electronic systems need to be designed with appropriate selection criteria, otherwise time will be sucked into the black hole of the computer screen.

Trust is important in a project team but not when it comes to progress information. With any form of delegation, particularly when work has been entrusted outside one's own organization, it is important to check that work is being performed as promised. Do not believe anyone when it comes to progress information, check it out, test that it works, go and see for yourself, and listen to good advice.

I was attending a conference on combined cycle power generation since it was relevant to my project. During the coffee break I bumped into the owner's chief accountant and during a chat he said, 'If I were you I would visit vendors X, Y & Z – they are clearly key to the project's success.' My first thought was, 'What does an accountant know about projects?' Reflecting on matters on the way back to the office I realized that he was right and organized visits accordingly. One visit remains fixed in my memory. I was treated like a visiting dignitary and, during a splendid lunch, I naturally asked, 'How is our equipment?' The response was, 'No problem – it is being assembled,' and before I realized it, I was being eased through a farewell process. Fortunately, through a fog of alcohol, my brain kicked in and I said, 'Whilst I am here I would like to see our equipment, after all that is what I came to do.' 'Oh! You don't want to get dirty,' was the response. I insisted, and we went to the assembly shop. It was not there. 'It must be in machining,' I was told. It was not there. Eventually we tracked it back through the manufacturing processes and found a casting wrapped up with grease in the stores! The managing director's embarrassment meant that our order then got the attention it deserved. Nowadays, whenever I hear overseas suppliers say 'no problem' I get worried!

DELIVERY

The delivery of goods is the responsibility of the transport department, the shipping department or the traffic department, depending on custom and practice in a particular company.

Whereas the delivery of goods is best left to the experts (true for any discipline!) the project manager may be able to provide assistance in the facilitation of the import or export of materials and equipment. Large items in particular, involving greater risk, require greater attention. Further, I see nothing wrong in asking the naive questions. 'What happens if...? Have you thought about...? How have you solved the...problem?'

The delivery of goods is a classic example of risk management. The large, high-profile, high-risk items get all the attention because the chances of failure increase significantly if anything about the planning of their delivery is substandard. Further, the impact and consequences of failure is great. Conversely, the more routine items may not get the attention to detail that is required, and mistakes will occur. Fortunately the impact of failure in these cases is considerably less.

The successful delivery of goods is not just dependent on the logistics involved but also on having the correct documentation. It can, therefore, be essential to have someone during the final stages of the project, who not only enjoys dotting the i's and crossing the t's, but is good at it.

The delivery of large items, particularly oversize loads, is a one-off opportunity that will not be repeated. It is, therefore, important to be prepared and to maximize the public relations opportunities involved. Publicizing the successes of a project enhances everyone's career. Even if the project is trying to keep a low profile it is important to take photographs for record purposes. Accidents can occur, there may be claims involved or they may be useful for future projects.

WHAT CAN GO WRONG

By the time the contract has been let the issues that will go wrong will mostly affect delivery. Common problems are:

- supplier bankruptcy;
- loss and damage to goods;
- strikes;
- transport issues.

All forms of transport will have problems: lorries get involved in accidents. I have an abiding memory of the road from Delhi to Agra literally littered at intervals with lorries on their sides due to overloading. It is not uncommon for ships to sink, and shipping is an area where the detail can catch one out. I was amazed at the bill for extras when I was responsible for chartering a vessel. Every chock or block of wood, every lashing or piece of rope was charged for.

Finally, aircraft have engine problems. One of my earlier projects involved importing aluminium pipe for an aircraft hydrant refuelling system. So that welding could start as early as possible it was decided to airfreight part of the order from North America. The transport aircraft developed engine problems and had to set down somewhere in the mid-Atlantic. Parts had to be flown out and, before repairs were completed, the crew had overrun their allotted flying hours. Consequently, a replacement crew had to be flown out. Could anything else go wrong? Well, yes. In order to make up some time we asked for permission for the flight to land at the airfield where the hydrant system was being installed. This was refused on the basis that it was a commercial flight but a military airbase. However, the flight became overdue and at around midnight they declared an emergency – I never did find out the reason. As a result they were allowed to land at the military airbase. In the meantime the shipment sent by sea had already arrived!

As project manager I could not understand why goods were taking so long to clear the docks and customs in Sri Lanka. So I asked our shipping people to arrange for me to visit the various warehouses, meet the people and build a relationship with the officials concerned. The enduring image I have is of a 3m high mass (approximately 7m by 4m) of old-fashioned Singer sewing machines mixed up with wooden boxes in every state of damage and disarray possible. Relationships were not the issue! Our problem was; is there a piece of our equipment underneath this horrendous heap and do we try and move it to find out? Since this situation was typical of all the goods in the various warehouses I left it to the procurement specialists who stationed someone permanently in each building. They then instituted an incentive scheme whereby anyone who found something we were looking for, but did not know about it, got a bonus payment. It worked very well.

As well as making sure that the goods have been packed to avoid loss or damage, it is important to receive them all together. Shipments involving multiple crates invariably arrive in the wrong order. The items that are required first arrive last, and usually much later, since they will most likely have been loaded onto a different vessel. Consequently, shipments involving letters of credit are usually annotated as 'partial shipments not allowed'. Unfortunately, on the project in Sri Lanka we got so used to ticking the box 'partial shipments not allowed' that it occurred on an order for 52 lorries. Now it was difficult enough shipping just a few lorries from southern India, let alone 52. I very soon learnt that shipping officials, and banks in particular, are not interested in project logic, but work to the precise wording of the documentation. Fortunately, on this occasion the delay in changing the paperwork did not affect the project schedule.

Another example that did not work out was when a project manager in Egypt decided to import new cars for use on one of our projects. Unfortunately, due to a lack of the correct paperwork, they got stuck in customs. Consequentially, storage charges were accumulated at an alarming rate. Customs then auctioned off the vehicles in order to pay for these storage costs. Replacement cars were bought locally!

A case that went to arbitration involved delivering a large vessel on a low loader on to a roll on-roll off ship. The Incoterms were Free On Board (known as FOB). The low loader duly drove into the ship and the vessel was offloaded. However, the client got the bill for the cost of offloading the vessel from the low loader once it was on the ship. The client objected since the delivery terms were free on board, and clearly they were not 'free'. Considering that the Incoterms define the responsibilities and liabilities for FOB as changing when the goods 'pass the ship's rail' the charges were correctly allocated. The Arbitrator, in spite of this, weakly decided on a 50/50 split of the costs between the supplier and the client.

Sod's Law says that if something can go wrong it will. So, obviously there are a hundred and one other things that can go wrong!

DELIVERY DIMENSIONS

A small 25mm connection can be very expensive when it is the deciding dimension for a wooden crate and shipping costs are determined by volume.

Special deliveries, due to the size of the equipment involved, can be significant projects in their own right. The delivery route will require meticulous planning and, in the UK, might involve getting approvals from up to a dozen statutory authorities. The route will require a detailed survey. Numerous obstructions will need to be removed and eventually replaced. Roads and bridges may need reinforcement. Thus, the initial decision to deliver in large modules, in order to save installation costs, will need to be balanced against these additional problems. Since oversize loads will require a police escort they will need to be booked well in advance. Unfortunately, the police have the ultimate authority during the journey and can change the route, if they think it is necessary, for whatever reason. On one occasion a large vessel, complete with all its attachments and finishes was re-routed and, as a result, got stuck under a bridge. It then had to be returned to the supplier's factory for repair, resulting in delays on site.

On another occasion, it was decided that it was logical to use a local power station's private dock to unload a large boiler furnace for our own project not that far away. A specialist heavy lift and transport company was used for the delivery and the furnace was unloaded from the ship onto a barge for transfer to the dockside. Unfortunately, when it came to unloading from the barge it could not be lifted off, regardless of the state of the tide. The dimension from under the crane hook to ground level was less than the sum of the height of the furnace and the slings and lifting tackle required. The equipment was left on the barge overnight. To everyone's surprise it was sitting on the dockside the following morning as intended. My curiosity was obviously aroused and I started to ask, 'How did you manage to do it?' I was quietly told, 'Don't ask.' Naturally, I eventually found out as discreetly as possible and never mentioned it again. Someone had used their judgement and experience and taken a calculated risk by removing the spreader beam that was part of the lifting tackle. If this had been done in the full glare of publicity and the senior management who were present during the day, it would never have been sanctioned.

In order to minimize schedule delays, lost and severely damaged goods should be reordered immediately. The loss of, or damage to, goods should automatically involve the implementation of a claim on the project insurance policy.

15 *The Loose Ends*

ADMINISTERING THE ORDER OR CONTRACT

Provided the right quality assurance checks were made before the vendor was selected, it is entirely practical to let an order for materials run its course. All that may be necessary is to check that the right test certificates are produced.

An order for equipment will require more checking that the termination points are correct together with witnessing of factory tests (Factory Acceptance Test, also known as FAT). For example a 'string test', that is putting all of the components of rotating machinery together in a line and checking that they operate satisfactorily as a single unit. Further, even though the order may be for a fixed price, there is likely to be a requirement for reimbursable vendor servicemen on a daily rate basis. It will, therefore, be important to check that all activities which should have been completed in the factory, on a fixed price basis, are actually complete. You do not want work left undone so that it can be completed during the installation process, at your cost, on a reimbursable basis.

When it comes to a contract for fabrication, installation or construction services, life is much more complicated. It is in this environment that the orders for materials and equipment become interdependent with the contract for services. Is the contractor managing the services ready to receive the materials or equipment? Do they have the necessary storage space and offloading machinery available? Are the foundations ready for the equipment to be placed in its final position?

Regardless of the relationship with a supplier or, more particularly a contractor, good management practice requires that we check up that they are performing in accordance with the contract. Consequently, administrative arrangements should have been agreed and incorporated into the contract for:

- the communication processes to be used;
- the administrative systems and procedures to be used;
- information, data or documents that need to be approved;
- approvals required for alterations to the project personnel;
- approvals required for the hiring, purchasing or use of additional resources;
- information and data to be provided by the client;
- free issue materials, equipment and facilities to be provided by the client or owner;
- progress meetings;
- frequency and type of progress reports;
- change order procedure;

- a procedure covering the use of vendor servicemen;
- inspections and tests that need to be witnessed;
- final acceptance or handover procedure;
- payment procedure.

Most importantly, however, the client must make sure that they do not let the supplier or contractor off the hook by failing to perform themselves. The contractor often relies on the purchaser failing to fulfil their obligations during the project execution phases. The client must make sure that documents, data, access, free issue materials and so on, that they agreed to supply, are made available on time. For example, '[The contractor][1] has said that problems arose because Treasury ministers denied them the time to test the [IT] system properly.' Further, for any form of process plant, the input conditions to a facility at the end of a project are often very different to those specified when the contract was awarded.

COMMISSIONING, SETTING TO WORK, START-UP

In order to achieve a successful project it is necessary to involve the users in the earlier process. This cardinal rule of project management is most clearly applied at the commissioning stage when the users are the operators of the equipment or facility. The following extract from a magazine[2] demonstrates this clearly:

> So why don't you see all these exciting new aircraft flying around at Boscombe Down? Well, because the way we work these days is to embed our trials officers and test pilots with the companies who are producing the new aircraft. We work as integrated teams, alongside the manufacturer, to try and shorten the time it takes to get new aircraft onto the front line. So whereas in years gone by most aircraft would have been tested by the manufacturers then sent to Boscombe Down for us to get our hands on, we try to do everything in one go now, working alongside the contractor at his factory. This not only speeds things up, but also allows our military perspective to be fed into new programmes at a very early stage – usually long before the aircraft actually flies. And of course the earlier you can influence the design, the more likely you are to get what you want at the end of the project. Our military, operational perspective is vital in ensuring that we, the three services, get equipment that will actually do the job it's being bought for.

Once a facility introduces live data or process fluids the consequences that flow from mistakes escalate considerably. It is, therefore, the owner that has to accept the financial and insurance risks involved. At one stage in my career I had a programme of projects that involved installing aircraft refuelling facilities on all the United States Air Force and NATO bases in the UK. The most interesting project was the design and installation of a hydrant refuelling facility for the Gallaxy aircraft – at that time the largest airplane in the world. Once construction and installation works were complete the system was filled with jet fuel and appropriate pressure tests carried out. However, the final test was refuelling an aircraft. Naturally I enjoyed asking the owner's representatives to provide a Gallaxy aircraft so that the system could complete its final test. Of course, the actual refuelling was carried out by United States Air Force personnel and there were also a lot of other anxious officials watching to see that everything was performed and operated in the correct manner!

---

1   'Revenue could sue tax credit IT supplier', by Nicholas Timmins and Scheherazade Daneshkhu. *Financial Times*, October 11, 2005.

2   'The view from my office' by Gp Capt Dave Best, Chief Test Pilot at Boscombe Down, *Boscombe Bulletin*, Winter 2005.

COMPLETION AND CLAIMS

The purchase order or contract cannot be considered as complete until the final payment has been made and any performance or retention bonds returned. Both of these activities may need the client to take the initiative. Whilst the final payment is dependent on receiving an invoice from a supplier or contractor, it may be necessary to chase them up in order to get an invoice! Bonds cost money as long as they have not been returned to the issuing bank. Consequently, the client should return them to the supplier or contractor as soon as the objective, for which they were issued, has been satisfied.

Unfortunately, it will probably not be as clear-cut and straightforward as described. Just when you think the order or contract is complete the supplier, or more likely, the contractor will submit a claim for additional costs.

There are a few basic rules for resolving claims that are raised at the end of a project. If the claim is of a major nature it is important for the project manager to recognize that they need to appoint someone independent from the project to act as claim team leader. The project manager and project team will be too keen to defend their previous behaviour to be sufficiently objective and dispassionate.

Firstly, you must insist on addressing the contract. Ask the contracting party raising the claim, 'Where does it say in the contract that I must pay you for this item?' Secondly, insist on supporting factual data. Check these data back to their source. Consequently, it is vital that the client maintains good project records right from the launch of the project. If matters arrive at this stage you will wish that you started maintaining your project records as if you had anticipated ending up in a court of law.

I remember one occasion when a contractor made a substantial claim for additional compensation. Their claim was based on the principle that it had rained more than it was reasonable for them to have foreseen at the tender stage. The supporting documentation that they provided from The Meteorological Office was not in dispute. In spite of this, the project team checked the data with the local meteorological office. The data proved to be correct. Yes, it had rained more than was normal, but only at night when no one was working!

Once the contractual issues and data are addressed, approximately two-thirds of the claim tends to evaporate. There are usually more than two opinions to the interpretation of what remains of the claim. Both parties will have failed to perform strictly in accordance with the contract and you, the client, will probably not have administered the contract as well as you would have liked. Consequently, it will probably be most sensible and practical to negotiate a deal of some form. Management do not want to be involved in lengthy contractual arguments and good project managers are needed for the next project.

SMALLER PROJECTS

Quite often people tell me, 'Everything you have explained is all very well for large projects. What about small projects?' The following real project was being undertaken at the same time that this chapter was being written. It is presented in order to answer the question about smaller projects and as a summary of the whole process described in this book. It may seem to be rather a long story but the process involved should not be compromised.

When my sister needed to redo the long pitched roof of her house in Burgundy she asked me to help her manage the project. This was a large project for her, but small in contracting terms, as well as being a new area of technology for both my sister and me! A small family firm (a father trying to transfer a business to his thirty-something son) had been suggested for the work, but it was deemed too insubstantial to consider further. Consequently, she surveyed the marketplace by observing the companies that were performing work in the area and obtained four or five estimates/quotations. In French law an estimate is the same as a quotation, a *devis* and, if accepted, enforceable as a contract. The pricing structure gave a good indication of the norms for the various materials involved, and the cost of the rather special Burgundian small tiles was checked with the local builders' merchants. You change the exterior appearance of properties in France at your peril – you can do what you like in the interior!

Unfortunately, prices from the large companies were far too high. So, after checking up on the work quality of the recommended small firm, by visiting a couple of buildings that they had re-roofed, they were asked to submit a *devis*. A fixed price would have been nice but the risk premium for unknown quantities would be too high. The roof structure would require new battons and some rafters would need replacing. In addition, extra rafters would also need to be inserted where the structure needed reinforcing. Further, as many as possible of the existing old tiles were to be recovered and reused, with additional second-hand old tiles purchased to complete one side of the roof. New old-style small tiles were then required for the second side of the roof. Consequently, the quotation was primarily made up of fixed rates, with a fixed price for the scaffolding and an estimated total cost. The payment terms they requested were 30 per cent at the start of the project, 30 per cent during the work, and 40 per cent upon completion. Apart from the first payment, the short schedule meant that payment would not exceed work progress.

During the early discussions the father only acted as an observer and made his son take the lead, and if he was addressed directly he would refer us back to his son. If we had only seen the son, negotiations would have been terminated. However, the father's determination to make his son perform professionally impressed us.

It was also felt that, apart from the overall cost, it seemed wrong to give the whole (new technology) project to one unknown small family company – too many risks in one basket. Although some of the schedule risk had, in effect, been transferred with the fixed price for the scaffolding.

Fortunately, the house structure naturally broke down into four equal sections and a revised *devis* was requested for one section. A motivating carrot was dangled by stating that phases two, three and four would be carried out in subsequent years. Two or three counter-offers ensued in order to get the scope precisely defined – work around chimneys, including a skylight, and so on.

Before any deal could be finalized we had to address the payment terms. I have always known that you do not pay builders up front, let alone before they appear on site. Television programmes about the blind faith of people who have lost substantial sums of money by paying builders in advance never cease to amaze me. I realized that if I asked for an advance payment bond his polite answer would have been, '*Vous faites une blague!*' (you are joking)! As a result, during my site visit to finalize the contract, I asked the father what was to stop them running off with the money. His offence at such an insult was, I believe, quite genuine. Nevertheless, he suggested that the down payment need not be handed over, in cheque form (laundering cash in France is a real problem now!) until they arrived on site.

Finally, to check that both parties understood the project the same way, my sister wrote out her understanding of the scope and price structure, and the contractor accepted it and signed it as a contract. In order to re-establish trust the down payment was made without further ado. I returned to the UK to receive a worried telephone call from my sister. Apparently the builder had told her that one of his clients had an emergency, and that they would not be able to start work until a week later than they had promised! Hey ho, I thought, here come the excuses, I should know better! However, a week later they appeared. The scaffolding went up and work was completed in the timeframe quoted and slightly under the estimated figure.

It would be nice to say that that was the end of the project, but it was not. The quality of the work at the back of the roof was perfect and conformed to the specification of using old tiles reclaimed from the front. However, much more importantly, the front was a mixture of new and old with unacceptably large areas of new, glossy, tiles. It eventually emerged that the builder was unable to find an appropriate source of old tiles. He steadfastly refused to replace the glossy new tiles with weathered second-hand ones until the client put her complaint into writing. On advice from the neighbours this had to be expressed in carefully worded diplomatic language so that the builder did not loose face. The client, having sourced the requisite tiles, offered to supply them free of charge and the rectification work was then carried out to produce a finished appearance that was just acceptable.

The other key issue was that the specification called for 5m on both sides of the roof. The builder claimed that he had completed 10m – 6m at the back and 4m at the front. The whole purpose was to make sure that the roof area around the two chimneys was made watertight. This had not been achieved. Fifty per cent of one had been dealt with and only 25 per cent of the second one. After another letter the remaining 50 per cent of the first one was rectified and tested with simulated rain from a garden hose. However, for some reason, they refused to do any work at all to the second chimney and a site visit was arranged to try and resolve the matter.

The contractor arrived in force with his father and adopted an aggressive stance. 'You should pay us for the work that has been done and only retain monies for the amount that is in dispute,' was their proposition. As soon as the client got out a tape measure, to get agreement on the facts of the work performed, there was a visible deflation in the body language of the contractor. Only 5.25m had been done at the back and, surprisingly, 4.5m at the front. It was clearly impractical to insist on the full specification with new battons, and so on, just for the shortfall of half a metre – it would probably do more damage than good - particularly as the half metre in question was in a reasonable state. Further, this section could be replaced, if it proved necessary, in phase two of the project. As the project manager I reminded the owner, my sister, that the objective was to be sure that the area around the chimney was watertight. After much argument, and when it was clear that no further payment would be forthcoming, the contractor agreed to seal around the chimney and replace the half-dozen tiles that were defective in the area in question. The contractor came, as agreed, the next day and the work was completed within a couple of hours.

It should then only have been a matter of making some minor adjustments to the quantities involved and settling the account. However, the client had some back charges. Two glass panes in the awning above the front had been cracked and would need replacing. 'Ah,' said the contractor, 'but I told your sister that I had fitted a length of galvanized tin (not in the specification) above the awning free of charge. She agreed that this cost could be offset against the cost of the glass.' I accepted that argument but pointed out that the glass still had to be fitted. Since he stated that he did not do glazing work we had to agree a cost for the fitting of the glass and adjust the account accordingly. Payment was made and the account signed as paid in full.

Was it a successful project? The scope and specification were just about acceptable. The objective had been achieved and relationships had been maintained. A qualified yes, to success. More importantly, experience had been acquired. The client will know where to make improvements in buying the same service next time. Would the client use them next time? Possibly, it depends on the marketplace.

POST PROJECT APPRAISAL AND LEARNING FOR THE

FUTURE

Firstly, big projects are different from small projects. Large projects need more formalized execution processes because the risks are greater. Similarly, large companies have more complex and integrated management processes.

Secondly, do not apply a big project attitude to a small project; the work practices will probably be inappropriate for more simplistic work on a small project. Similarly, do not apply a large company management philosophy to a small company; the more sophisticated procedures will make the project more expensive.

So, what did I learn from the above smaller project? Further, how do the issues relate, more generically, to differences between large projects and small projects?

- Regardless of the size or type of project; do not forget the objective or the purpose for which it was intended.
- The contract strategy was the same as if it were for a large project.
- The contract strategy whereby the client took the reimbursable quantities risk was validated by the outcome being slightly under budget.
- The unknown company and unproven performance of the small company was mitigated by recommendation and by visiting previous projects.
- The project risks were all reduced by breaking the project down into smaller chunks.
- The programme risk was greatly influenced by other projects and the limited resources of the smaller company.
- Schedule and productivity risk was mitigated by the implication of future work.
- Some schedule risk was also covered by the contractor assuming the risk for scaffolding.
- The risk analysis for a larger project is more focused at a higher level. Whereas on a smaller project it is possible to analyze the risks in more detail and to a greater depth.
- The need for comprehensive scope and specification documents is the same for large or small projects. Clearly, in this case, the specification was deficient. A sketch should have been produced and it would have been much better if it had used the key words; that the front and back 'should be identical'.
- The risk associated with the supply of project materials can be mitigated by the client assuming this risk. However, if the client purchases the materials and supplies them free of charge, there is then the different risk that the materials will get lost and damaged.
- The overall financial risk was mitigated by reducing the scope.
- The payment terms were not that different.
- The financial resources and cash flow problems for smaller companies are such that payments have to be made earlier and spread through the job more evenly than they need to be for larger companies.
- A commercial risk was taken regarding payment security, it was based on evaluating and trusting one individual. Perhaps this is a difference on smaller projects – trust in your own judgement has a higher profile. However, a significant proportion of the project cost (40 per cent) was retained until agreement was reached that the work was complete.
- It would be difficult, if not totally impractical, to apply the payment security issues of bonds and guarantees to small projects. A small company is unlikely to have the assets needed for a bank to be willing to issue a bond of any form. For the same reason a parent company guarantee is probably not worth the paper it would be written on.
- The tender evaluation is more personal and perhaps more subjective. Possible evaluation criteria being: price, quality, reliability and trust.

- The contract was awarded based upon an interview with a key member of management. The MD wanted the job.
- Assessing, evaluating and analyzing the character of the key personnel has a greater degree of importance on small projects. On large projects you can change personnel. On small projects you may need a greater degree of trust in individuals than you will feel comfortable with.
- When there are problems on a larger project it is always possible to bring in people with a new and more objective perspective by escalating matters to a more senior management level. On a smaller project with smaller contractors you are probably already dealing with the managing director.
- There are no fundamental differences between large and small projects when settling claims.
- The project management of smaller projects is possible with the project manager visiting when needed. However, it is essential to have a representative on site to record what happens, and to advise the project manager when a site visit is necessary.

Overall, my conclusions are that the main differences between large and small projects are all connected to matters relating to resources. Namely, management personnel and labour, equipment for performing the work, materials and financial resources.

An advantage with a smaller contractor – not necessarily a smaller project, is that you will most likely be dealing with the same people if you use them again. We often use the same company again because we feel confident that we know them. They have been through the learning curve, they understand our requirements better, we know their deficiencies, and where they need supporting. It is the relationship that matters. If you deal with the same company again and they are all new people, you might as well be dealing with a new company.

The last activity in acquiring goods and services is not solely related to the purchasing function but more to do with procurement as a whole. The project procurement manager must prepare a final report for the procurement activities, not only for inclusion in the project manager's historical report, but also for their own department's use. The final report is usually poorly performed for most projects. Time has run out, it is not exactly stimulating work, and the next assignment is on the horizon, which is much more exciting. Without analyzing what went well and what could be performed better the next time, the expertise of the organization will stagnate. This is particularly true of purchasing, whose capability is dependent on knowledge of the marketplace. Consequently, they probably perform this activity better than other functions. It is hoped that the checklist that follows will help the readers' performance.

PURCHASING PLAN AND CHECKLIST	Reference Number:	
Materials/Equipment or Services	PLANNED Date – Yes/No Option Chosen	ACTUAL Date Achieved
PREPARATION PHASE		
Approval authority to proceed		
Budget and schedule date agreed		
Risk analysis		
Contract strategy		
Market survey		
Preliminary long tender list		
Allocation of risk and type of contract		
Preliminary talks with potential tenderers		
Material requisition number		
Scope of work		
Drawings/sketches/plans		
Specification/standards/samples		
Key programme dates		
Critical path item(s)		
Delivery location		
Check meaning of key words		
Invite expressions of interests		
Pre-qualification		
Agree short project tender list		
Terms: standard or tailored		
Incoterms		
Payment terms specified or requested		
Inflation/CPA clause		
Insurance requirements		
TENDERING PHASE		
Enquiry method – open/selective		
Local/national/international		
Tender/proposal/quotation		
Verbal/written/sealed tender process		

PURCHASING PLAN AND CHECKLIST	Reference Number	
Materials/Equipment or Services	**PLANNED** **Date – Yes/No** **Option Chosen**	**ACTUAL** **Date Achieved**
Issue enquiry/invite tenders		
Durations for tender programme		
Date tenders due		
Tender clarification meeting		
Arrange site visits		
Arrange interviews		
Schedule Tender Review Board		
EVALUATION PHASE		
Tender analysis criteria		
Tender opening process		
Distribute copies to other departments		
Duration of evaluation period		
Required evaluation completion date		
Legal/commercial inputs		
Interview key personnel		
Negotiation 'must haves' identified		
'like to have & nice to have' issues		
Recommended supplier/contractor		
PM approval		
Management approval		
Owner approval		
CONTRACT PHASE		
Order placement date		
Order acknowledgement		
Enter on Material Status Report		
Debrief unsuccessful tenderers		
Owner/client inputs required		
Data to be provided		
Contract administration procedures		

PURCHASING PLAN AND CHECKLIST	Reference Number	
Materials/Equipment or Services	**PLANNED** **Date – Yes/No** **Option Chosen**	**ACTUAL** **Date Achieved**
Delivery arrangements agreed		
Contact names & telecom nos. established		
Free issue materials		
Vendor servicemen		
Inspection authority		
Inspection by installation personnel		
Commissioning and maintenance spares		
Capital spares		
Operating and maintenance manuals		
Claims settled		
Liquidated Damages		
Maintenance period		
Retention monies released		
Bonds returned		
Final account paid		
Historical reports from all functions		

Additional Sources and Contacts

APM, Association for Project Management. www.apm.org.uk
Their Body of Knowledge provides brief outlines of the subject areas together with references to recommended books and publications.

British Standards Institution BSI. www.bsi-uk.com
BS 0, is 'a standard for standards'. It sets out the way British Standards produces standards. It was revised and came into effect 1 January 2006. Standards can be searched for online at www.bsonline.bsi-global.com. The BSI Head Quarters is: BSI Group www.bsi-global.com.

www.ectenders.com.
Provides a service to promote contract opportunities issued by public authorities. There is useful information, however, you are required to be an accredited supplier to view the specific contract opportunities.

European Communities 'Practical Guide' explaining the contracting procedures applying to all external aid contracts: http://ec.europa.eu/europeaid/tender/practical_guide_2006/documents/new_prag_en_final.pdf
Please note the final part of the address: practical_guide_2006/documents/new_prag_enfinal.pdf

First Point Assessment Ltd. 7 Burbank Business Centre, Souterhead Road, Altens, Aberdeen AB12 3LF. Tel 01224 337500. www.fpal.co.uk
The interesting information is for registered users only.

Guide to the Preparation and Evaluation of Build-Own-Operate-Transfer Project by Tony Merna and N.J. Smith. Published by Asia Law & Practice Books, 1996 (out of print.) ISBN 9627708720.

Guidelines, Procurement under IBRD Loans and IDA Credits The World Bank, 1818 H Street, N.W., Washington, D.C. 20433. www.worldbank.org /procurement/resources/procguid-ev3.doc. ISBN 0-8213-3218-X.

Incoterms 2000. The ICC (International Chamber of Commerce) official rules for the interpretation of trade terms. Entry into force 1st January 2000. ICC Publishing SA. ISBN 92 842 1199 9.
www.iccwbo.org/incoterms takes you to the home page which will link to their 'preambles' from which specific terms can be accessed. There is a useful note on variants of Incoterms. However,

the site does not spell out the obligations of the Buyer and Seller. The Incoterms book has to be purchased.

International Organization for Standardization ISO. www.iso.org

International Trade Centre. www.intracen.org.

You are required to be a registered supplier. However, the 'Award procedure' provides a link with information.

Office of Government Commerce. www.ogc.gov.uk

Open the procurement section then click on Procurement Policy then select Procurement Policy and Application of EU Rules.

Official Journal of the European Union (OJEU)

www.ojec.com provides access to www.mytenders.org. 'myTenders is a market proven solution for the publication and management of tender notices.' (www.tendersdirect.co.uk is a tender tracking system for suppliers and contractors).

Partnering in Europe: Incentive Based Alliancing for Projects, edited by Bob Scott. Produced by the European Construction Institute (ECI). Published by Thomas Telford Ltd, 2001. ISBN 072772965-9

Index

Printed and bound by CPI Group (UK) Ltd, Croydon, CR0 4YY

18/10/2024

01776204-0003